The Sustainable University of the Future

Mariam Ali S A Al-Maadeed
Abdelaziz Bouras
Mohammed Al-Salem • Nathalie Younan
Editors

The Sustainable University of the Future

Reimagining Higher Education and Research

Editors
Mariam Ali S A Al-Maadeed
Research and Graduate Studies
Qatar University
Doha, Qatar

Abdelaziz Bouras
Research and Graduate Studies
Qatar University
Doha, Qatar

Mohammed Al-Salem
Office of Research Support
Qatar University
Doha, Qatar

Nathalie Younan
Research and Graduate Studies
Qatar University
Doha, Qatar

ISBN 978-3-031-20185-1   ISBN 978-3-031-20186-8   (eBook)
https://doi.org/10.1007/978-3-031-20186-8

© The Editor(s) (if applicable) and The Author(s), under exclusive license to Springer Nature Switzerland AG 2023

This work is subject to copyright. All rights are solely and exclusively licensed by the Publisher, whether the whole or part of the material is concerned, specifically the rights of translation, reprinting, reuse of illustrations, recitation, broadcasting, reproduction on microfilms or in any other physical way, and transmission or information storage and retrieval, electronic adaptation, computer software, or by similar or dissimilar methodology now known or hereafter developed.

The use of general descriptive names, registered names, trademarks, service marks, etc. in this publication does not imply, even in the absence of a specific statement, that such names are exempt from the relevant protective laws and regulations and therefore free for general use.

The publisher, the authors, and the editors are safe to assume that the advice and information in this book are believed to be true and accurate at the date of publication. Neither the publisher nor the authors or the editors give a warranty, expressed or implied, with respect to the material contained herein or for any errors or omissions that may have been made. The publisher remains neutral with regard to jurisdictional claims in published maps and institutional affiliations.

This Springer imprint is published by the registered company Springer Nature Switzerland AG
The registered company address is: Gewerbestrasse 11, 6330 Cham, Switzerland

Foreword by Prof. Pam Fredman

Knowledge and understanding, acquired thanks to quality higher education teaching and learning, research and community engagement are essential for our societies to thrive. The coronavirus pandemic has come with a strong call for true transformation of the sector, and many HEIs have been faced with challenges to deliver on their promise due to lack of capacity to move to remote teaching and learning and innovative ways to carry out research under unforeseen circumstances. This has also called for new teaching pedagogies to be developed. Universities have put considerable emphasis on developing new, more innovative models, and on challenging their own boundaries.

Moreover, as also underlined by the UNESCO Futures of Education Report, a new "social contract for education" has been called for as some of the past educational shortcomings and in particular the social inclusion imperative need to be addressed. Research and innovation are key cornerstones of such a new social contract for education. Higher education needs to be more student centered and at the same time better connected to the grand sustainable development challenges of our time.

The digital transformation of teaching and learning, research, and even community engagement can help find new solutions to the challenges the sector and society face. It offers new opportunities to connect people and disciplines across the globe and, when done well, offers unique opportunities to develop novel solutions to issues that call for original approaches. Technology can help support the new learning processes called for, and when used appropriately, it can help tailor to the needs of individual learners while also offering to connect them to larger groups of learners. This calls for mechanisms that will help the academic staff to acquire new competencies for them to use to develop innovative learning opportunities.

This collection of chapters published in this book provides a true contribution to the existing body of knowledge and is of special interest to those who are interested in the transformation of the university and its development in the future.

President of the International Association of Universities (IAU)
Paris, France

Pam Fredman

Foreword by Prof. Sheikha Abdulla Al-Misnad

Higher education has a long history of a permanent capacity for adaptation and innovation. However, at this critical historical moment, where the world is at crossroads, reimagining the future of higher education is more essential than ever. At a smaller timescale, the world of education experienced challenges during the recent pandemic period that no one could have imagined. The pandemic had taught us that change at scale is possible, and ongoing organizational, technological, and pedagogical changes have been initiated by higher education institutions. Indeed, higher education needs to be more open, more inclusive, and more resilient. A roadmap for a new era of education is necessary to go beyond the challenges of our current systems toward a sustainable future. It requires our most creative capacities, with the imagination and innovation needed.

This book provides opportunities to the members of academic institutions and governmental agencies to better understand the current mutations and explore their implications on higher education and on how to reimagine the university of tomorrow.

Former President of Qatar University (QU) Sheikha Abdulla Al-Misnad
Doha, Qatar

Preface

This book titled "Sustainable University of the Future: Re-imagining Research and Higher Education" represents a set of important-needed topics to address nowadays. Such an inspiring title discusses the requirements imposed by the rapid changes on higher education institutions that have pushed them to reconsider their traditional policies and plans and adopt more sustainable and modern programs and tools, which keep pace with transformations and anticipate the next challenges. In addition, it expands opportunities and options for faculty members and students, transcending boundaries and distances.

The "Sustainable University of the Future" must play its role in sustainable development, despite the current regional and international challenges. The universities in the new era need to be transformed to lead and adopt the new changes in the world.

Furthermore, the adventure of Industry 4.0 is already assuming its rhythm in today's business environment and agendas. Companies are deeply changing their working processes, which pose major challenges as they require proactive adaptation to the digital culture, particularly regarding their workforce' skills and competencies. Hence, it is critical that industries and universities join efforts and take an active role in preparing the next workforce and, simultaneously, support the current once through lifelong learning and continuous training. This should be done by bringing together actors such as businesses, universities, public authorities, and students to address the existing gap in higher education and cocreate innovative and multidisciplinary solutions adjusted to the current and upcoming challenges of the digital era.

The book has twelve chapters, covering some of nowadays needs:

The first chapter investigates the various methods of localizing sustainability in achieving the SDGs in the university context with a case study from the Gulf region. It shows how a higher degree of transition toward sustainability may potentially allow the university to adopt a more proactive role in tackling sustainability issues and putting SDGs into practice on the societal level.

Chapter "Leveraging Eruptive Technologies and Systems Thinking Approach at Higher Education Institutions" focuses on predictive learning analytics and systems thinking approaches at HEIs, to create sustainable future universities. It suggests that adopting a perspective system to problem situations provides a more helpful framework for representing the real world than the reductionist, discipline-based one. The chapter demonstrates how these techniques could unlock actionable intelligence from statistical analysis of student data combined with other traditional data sources like demographics or academic success to make actionable decisions regarding students, learning and course design, and institutions.

Chapter "Reimaging Continuing Professional Development in Higher Education—Toward Sustainability" states that the importance of Continuous Professional Development (CPD) in higher education with respect to promoting sustainable education lies in the concept of collaboration. The use of CPD transformational models such as implying action research and transformation model within HEIs would promote sustainability in terms of improved teaching and research. It states that the transformational model of CPD is a gist of all other models of CPD, as it provides power to the teachers so that they can determine and exercise their own learning pathways.

Chapter "Re-designing Higher Education for Mindfulness: Conceptualization and Communication" deals with redesigning higher education for mindfulness and contributes to the literature by forming new models of mindful consumption definitions of current generations. As its findings may expand to the following generations, it proposes suggestions to redesign higher education in order to comply with the needs of the youth to overcome the barriers against mindfulness and to realize behavior change.

Chapter "Preparing Future-Fit Leaders for the Sustainable Development Era" presents a framework, a model, and an application example from practice to guide HEIs implementation of education for sustainable development. It introduces three action principles that would guide curriculum development and implementation—cultivating capabilities, curating knowledge, and connecting networks—and promotes adaptation and collaboration that underpin individual and collective transformations.

Chapter "Changes Required in Education to Prepare Students for the Future" deals with changes required in education to prepare students for the future. It highlights potential uses for technology in early and mid-level education, and how technology-focused instruction can replace or be given alongside the existing educational framework. It also explains that by focusing on what society needs and wants as we transition to smart cities, we can adjust the educational system to produce graduates with education in these subjects. It concludes by offering solutions for maintaining relevancy and sustainability in the proposed educational projects.

Chapter "Humanising Higher Education: University of the Future" explains that it is essential to bring about a balance between the utilitarian and mechanistic structure aimed at employment, innovation, high-tech, and market value and a structure characterized by sympathy, empathy, compassion, and values that humanize life. Likewise, the key performance indicators (KPIs) must be rightly balanced with KIP (key intangible performance), things that cannot be measured in the conventional and structured way. It is stated that there can be nothing more important to civilization than to bring the soul back into education as the interaction between human beings and advanced technology shares the global stage for the better of humanity and for mercy to all.

Chapter "Social Science and Humanities in Future University" looks into the debate regarding social sciences and humanities in future universities. Through the appraisal of existing arguments, contentions, and challenges, the chapter studies how social sciences and humanities can survive the lull that is being experienced in academia. The challenge of image faced by social sciences, the issue of job opportunities, and the actual impact experienced and exerted by social sciences are some of the main issues highlighted in the course of the chapter.

Chapter "Impact of the Industry 4.0 on Higher Education" highlights the impact of Industry 4.0, not only on technology advancements but also on people in the workforce. It insists on how HEIs must properly train students in order to better adapt to such changes and on the fact that the innovation and entrepreneurship management skills of the students should be developed through the restructuring of future university programs. Finally, it shows how a new leadership mindset and governance can be crucial to tackle the challenges associated with education 4.0 and harness the future benefits of IR 4.0 while protecting humanity from potential threats carried by technology.

Chapter "University-Industry Transformation and Convergence to Better Collaboration: Case Study in Turkish Food Sector" shows, through a Turkish case study, that university-industry cooperation in higher education is not a choice anymore but a necessity for the future of the universities. This is not an issue to be overlooked by simple legislative barriers or formal practices. It explains how the parties should use a definite and clear will to create an effective and functional institutional cooperation structure. Technology centers stand out as extremely efficient and indispensable applications that need to be developed.

Chapter "Sustainable Development Goals Through Interdisciplinary Education: Common Core Curriculum at University of Hong Kong" presents, through a case study in Hong Kong, how academics can utilize SDGs as broad, holistic, and flexible frameworks to frame, contextualize, and comprehend the series of issues, challenges, and complexities. It also shows the implications regarding the incorporation of SDGs into the university curriculum. Such implications are the prioritization of SDGs in accordance with the unique contexts and features of the curriculum, the consideration of their interlinkages in the planning and designing stages of curriculum and courses, and the need for some realistic and concrete means to bridge these goals with the curriculum and courses.

Chapter "Implementing the Sustainable Development Goals (SDGs) in Higher Education Institutions: A Case Study from the American University of Beirut, Lebanon" describes the steps taken by a university in Lebanon to formulate a strategic plan for the implementation of SDGs in domains such as research and education in the age of Industry 4.0, sociocultural impact operations, governance, and external leadership. The implementation components follow steps such as rallying forces around SDGs and launching the vision, connecting the university to SDG networks, promoting SDGs initiatives on campus, reporting and increasing the university ranking on SDGs, and dreaming the future!

This book is an excellent resource for anyone involved in research and higher education systems. It will be essential for academics, researchers, education specialists, students, governmental agencies, and industry sector, who want to understand the evolution of research and higher education systems around the world, and how it will change the society.

We would like to thank the authors and the reviewers for their valued contributions and efforts.

Our thanks also go to Qatar University Research Forum and to the ANDD Academic Network for Development Dialogue (in cooperation with ESCWA United Nations Economic and Social Commission for Western Asian and ACUNS Academic Council on the United Nations System) for their continuous support. This book has been presented during a roundtable at ACUNS 2022 in Geneva, for which we thank all the participants for their valuable contribution.

Doha, Qatar

Mariam Ali S A Al-Maadeed
Abdelaziz Bouras
Mohammed Al-Salem
Nathalie Younan

Contents

**Universities of the Future as Catalysts for Change:
Using the Sustainable Development Goals to Reframe
Sustainability – Qatar University as a Case Study** 1
Esmat Zaidan, Emna Belkhiria, and Cesar Wazen

**Leveraging Disruptive Technologies and Systems
Thinking Approach at Higher Education Institutions.** 25
Mhlambululi Mafu

**Reimaging Continuing Professional Development
in Higher Education – Toward Sustainability** 43
Saba Mansoor Qadhi and Haya Al-Thani

**Re-designing Higher Education for Mindfulness:
Conceptualization and Communication** 63
Damla Aktan and Melike Demirbağ Kaplan

Preparing Future-Fit Leaders for the Sustainable Development Era 83
Sarmad Khan

Changes Required in Education to Prepare Students for the Future 107
Lobna A. Okashah, Akram Hamid, Jiwon Kim, and Ethan Rubin

Humanising Higher Education: University of the Future. 119
Dzulkifli Abdul Razak and Abdul Rashid Moten

Social Science and Humanities in Future University. 133
Mahjoob Zweiri and Lakshmi Venugopal Menon

Impact of the Industry 4.0 on Higher Education. 149
Thafar Almela

**University-Industry Transformation and Convergence
to Better Collaboration: Case Study in Turkish Food Sector** 165
Ece İpekoğlu

Sustainable Development Goals Through Interdisciplinary Education: Common Core Curriculum at University of Hong Kong 177
Adrian LAM Man Ho

Implementing the Sustainable Development Goals (SDGs) in Higher Education Institutions: A Case Study from the American University of Beirut, Lebanon 199
Mirella Aoun, Rami Elhusseini, and Rabi Mohtar

Index. 217

About the Editors

Mariam Ali S A Al-Maadeed is Vice-President for Research and Graduate Studies and Professor of Physics and Materials Science at Qatar University. Prof. Mariam has been leading the research and graduate studies strategies, programs, and developments since 2016. She played a major role in developing research and graduate programs, as well as fostering and strengthening national and international relations within this field. She is also the holder of the Leadership Excellence Award for Women from the conference and exhibition of engineering for the Middle East in 2015. Prof. Mariam is currently chairing the Academic Network for Development Dialogue, a new UN initiative related to implementing the SDGs within universities of the region and leading the initiative of the university transformation.

Abdelaziz Bouras is Professor in the Computer Science and Engineering Department of the College of Engineering at Qatar University, and Acting Director of the Office of Research Support. He was previously Professor "Exceptional Class" at Lumiere University of Lyon, France. He has managed several international projects (in Europe and the Gulf region) and coordinated several Erasmus-Mundus programs between EU and East Asia. He has published many research papers in refereed journals/conferences and published several books, one of them related to the university-industry collaboration within the Industry 4.0 context. He is currently chairing an IFIP International Federation of Information Processing working group related to life-cycle management and SDGs.

Mohammed Al-Salem is Professor in the College of Engineering at Qatar University and former Director of Research Support. He initiated several research and innovation programs that had a great impact on the transformation of Qatar University. He also held the position of Head of Department of Mechanical and Industrial Engineering for several years. During his term, Prof. Mohammed introduced innovative pedagogical tools and platforms contributing to the digital transformation of his department.

Nathalie Younan is a project manager at the Vice-Pesident Office for Research and Graduate Studies. She deals with research and education innovation and leads the annual Research Forum on the University of the Future. Nathalie has a notorious experience related to higher education institutions' transformation models and has built a solid international network of university decision-makers on the topic. She holds Master and Bachelor degrees from EPFL Ecole Polytechnique of Lausanne, Switzerland.

Universities of the Future as Catalysts for Change: Using the Sustainable Development Goals to Reframe Sustainability – Qatar University as a Case Study

Esmat Zaidan, Emna Belkhiria, and Cesar Wazen

Abbreviations

ANDD	Academic Network for Development Dialogue
CCE	Community Service and Continuing Education Center
DESD	Decade for Education for Sustainable Development
ESCWA	Economic and Social Commission for Western Asia
GCC	Gulf Cooperation Council
HEIs	Higher Education Institutions
ISOCARP	The International Society of City and Regional Planners
MDGs	Millennium Development Goals
MIA	Museum of Islamic Art
MOU	Memorandum of Understanding
MRV	Measuring, Reporting and Verifying
QICCA	Qatar International Center for Conciliation and Arbitration
QNDS	Qatar National Development Strategy
QNV	Qatar National Vision
QSTP	Qatar Science and Technology Park
QU	Qatar University
SDGs	Sustainable Development Goals

E. Zaidan
College of Public Policy, Hamad Bin, Khalifa University,
Doha, Qatar

E. Belkhiria (✉)
Institutional Research and Effectiveness Department, Chief Strategy and Development Office,
Qatar University, Doha, Qatar
e-mail: emna@qu.edu.qa

C. Wazen
Office of International Affairs, Qatar University, Doha, Qatar

© The Author(s), under exclusive license to Springer Nature Switzerland AG 2023
M. Ali S A Al-Maadeed et al. (eds.), *The Sustainable University of the Future*,
https://doi.org/10.1007/978-3-031-20186-8_1

SESRI	Social and Economic Survey Research Institute
SMRC	Sidra Medical and Research Center
SSD2019	Science for Sustainable Development 2019
THE	Times Higher Education
UAV	Unmanned Aerial Vehicle
UN	United Nations
UNESCO	United Nations Educational, Scientific and Cultural Organization

1 Introduction

Human activities have given rise to significant perturbations of natural processes, further causing harm to socioecological systems and inevitably contributing to the current environmental crisis. With the growing environmental, economic, and social issues globally, the world is quickly approaching a tipping point that poses a threat to planetary boundaries. Over the next few decades, the Earth's natural ecological systems could reach a "point of no return" if necessary measures are not undertaken to curb greenhouse gas emissions and promote and foster environmental stewardship on a global level. The Sustainable Development Goals (SDGs) of Agenda 2030 were established and adopted by 193 UN member states in 2015 to establish a plan of action for the next 15 years. The global sustainability agenda consists of a set of goals and aims to achieve sustainable development on a global scale by 2030 with the key targets of poverty eradication, preservation of the planet, and establishing prosperity for societies worldwide (United Nations Sustainable Development Summit, 2015).

Sustainability has become a national vision for most countries in a world increasingly characterized by global impacts, modern risks, and unstable, uncertain, and complicated conditions [2, 39]. The SDGs play the role of a compass by providing a transformative development framework for the transition to inclusive and sustainable societies and to building effective, inclusive, and accountable institutions that ensure access to education and information.

Considering the main consumers and creators of information concerning sustainable development issues of their countries and beyond, higher education institutions (HEIs) are catalysts for a more equitable society and sustainable development in their pursuit of sustainability. Consequently, the contribution of HEIs to sustainable development rooted in the national visions of their countries and in the global sustainability agenda largely depends on the alignment of their policies and strategies with the national visions of their countries and their adaptation of the SDGs strategically [40]. Doing so enable them to be more responsive to the needs of their immediate communities with respect to the implementation of the development plans, national visions, and the SDGs. To attain such alignment and to achieve a transformative agenda, pedagogic innovation and research should meet the priorities of the wider stakeholder community in its quest for sustainability. Thus, HEIs play a vital role as change agents considering the need to connect their communities with the external communities that they serve and with which they interact locally, nationally, and internationally. On the one hand, a HEI campus, as a major community within cities, possesses unique characteristics associated with energy, the environment, society, and the economy. Due to their dynamic energy use and available onsite resources, such campuses can interact with their surrounding environment and work toward achieving the goals of the key performance indicators of smart transformations toward a sustainable post-carbon agenda. On the other hand, being rooted at the local level and connected at the global level, HEIs are leading scientific and technological advances based on a global research agenda to educate future leaders and citizens and to deliver the knowledge needed for the successful implementation of the SDGs.

This chapter aims to explore the transformative role of HEIs in achieving the SDGs in academia and research to establish a sustainability culture that will influence the next generation. It provides insights into the policies and strategies that a university could adopt to better align with the SDGs and to link the university community with the external communities with which the university interacts and that it serves. Furthermore, the chapter reflects on and draws conclusions about the voyage toward sustainability at a strategic level, using the SDG framework toward delivering sustainable development and creating a balance between community-integrated smart campuses and teaching and learning strategies to contribute to the transition to sustainability. The chapter uses Qatar University (QU) as a case study.

2 New Role of Higher Education Institutions

Studies related to universities and colleges as organizations have developed over the past five decades [45, 47, 49]. The study of institutions has been one of the concerns of social research, in which institutions are seen as a functional necessity of complex societies [13, 51]. The traditional purpose of higher education, per Humboldt's idea of a university [2], was to pursue impartial truth via research and teaching [29]. Universities were "places which sought knowledge for its own sake, as part of an educational ideal which molded the personality of the cultivated man" [28]. Some expansion of the core missions of universities can be observed, with attentiveness to community and growth of knowledge economy, public engagement, and national development added to research and teaching [11, 27]. Another feature of the Humboldtian model was that universities must maintain a distance from the outer world to serve as society's critical conscience [3]. The role of HEIs has expanded over time, along with a growing body of knowledge that has pushed HEIs to participate in the sustainability agenda by integrating sustainability into their processes, vision, and systems [5, 41].

Currently, sustainability is observed as part of universities' missions, program curricula, and institutional policies, in addition to outreach and engagement initiatives [37]. Other practices include research, collaboration, data reporting and analysis, and, in some cases, the reduction of greenhouse gas emissions [5].

Different from any type of organization, HEIs, with their dedication and aim to increase knowledge, skills, awareness, and values, represent one of the most important vehicles toward sustainability.

However, this role requires a dramatic and systematic change at all levels, with a need for awareness and shared responsibility [4]. Investigating the new role of HEIs in achieving sustainability given the complex factors that surround the higher education ecosystem's interactions will facilitate understanding of the trends and patterns in sustainable development played by HEIs.

3 Sustainability and Higher Education: Strategizing Sustainability and Localizing the SDGs.

The sustainability transition implies an undergoing process of fundamental change in production and consumption patterns in reaction to imminent sustainability challenges. The water and energy sectors are confronted with fundamental restructuring in light of increased resource depletion, emerging risks, and international sustainability goals [1]. Some examples of the overarching sustainability challenges include depletion of groundwater resources, air pollution, greenhouse gas emissions, management of risks and accidents in the Gulf waters, and impacts of energy and water production on marine ecosystems, reducing large consumption footprints or improving water use efficiency in urban settings. In response to these environmental and sustainability challenges, key sectors, such as water and energy, are undergoing technological renewal to achieve a greater degree of sustainability [40]. Today, sustainability is extremely important, with the use of SDGs as a key instrument to direct action in a world characterized by unstable, uncertain, and complex conditions. Universities also play a crucial role as drivers of change.

In contrast, achieving the targets of the Agenda 2030 goals requires a strong intent from those involved in the community sphere to pave the way toward a sustainable future by changing the current trends of development. As such, society can play a key role in pressuring the government and offsetting corporate interests, which are providing a refuge to the status quo. Consequently, it is of the utmost importance to effectively educate the public on the underlying factors and consequences of the environmental crisis, as well as to formulate mechanisms to effectively manage corresponding challenges [21]. Since HEIs are locally embedded into their communities, as well as globally connected, through close collaboration with faculty, students, alumni, and the surrounding stakeholder community, they can provide significant opportunities to achieve the targets of the SDGs [12, 31]. Universities play a vital role in the effort to achieve the targets of the SDGs. Specifically, HEIs can save students from being burdened by a feeling of detachment and hopelessness arising from the current unprecedented situation, while encouraging the acquisition of key skills and ideals of harmony between nature and human beings [26].

Higher education can provide innovative new solutions for the world, addressing the key issues of our time and age as revealed in the SDGs (United Nations Sustainable Development Summit, 2015; [31]). In fact, since the Rio Summit in 1992, HEIs have been working on sustainability-related challenges through the implementation of Agenda 21 [26]. Being at the center of technological and scientific advances, global research initiatives, and the creation of professionals and future leaders, universities provide key expertise and knowledge in every sector and function as anchors in the communities that they serve on both the national and international levels. These agendas provide a plethora of opportunities for researchers and HEIs to contribute both academically and practically to the various overlapping aims of these agendas. A mixture of technical, scientific, administrative, and political support is required for every nation interested in addressing the issue of SDG localization with respect to national development priorities. A collective and shared approach is required to stay within the territory of the SDGs' all-encompassing and bottom-up approach. Of key importance is the idea that the SDGs represent an agenda for development that should be adopted and implemented by both developed and developing nations alike [21].

Education for sustainability became a key concept after the United Nations' Decade for Education for Sustainable Development (DESD) 2004–2015. This concept was followed by massive efforts from various countries to incorporate DESD into their policies and processes. In addition, additional attention was paid to higher education as a key player in promoting sustainable development [43]. Utama et al. [34] defined five strategies that might enable key initiatives that could further fast-track the process of localization of SDGs for HEIs. These strategies encompass improvements in HEI quality, equity, environment, research and innovation, and partnerships on both the global and local levels. Aside from the broader SDG agenda, there is a particular SDG (SDG 4) that is directed at HEIs [21]. This particular SDG consists of ten targets and several corresponding indicators.

Universities can help to pave the way for a shift toward a more equitable society and a better world by implementing the SDGs at a strategic level in search of sustainability and as a mechanism for forming close ties of higher education with business, industry, community partners, health care, and entrepreneurs [12]. Through its emphasis on education, Agenda 2030 can open doors for a transformative shift toward sustainability for HEIs [26]. Agenda 2030 comprises an SDG dedicated exclusively to education (*SDG 4: Quality Education*) and a corresponding target that defines education for sustainable development (target 4.7) while emphasizing the key engagement of sustainability education with the other 16 SDGs [32].

This emphasis allows HEIs to communicate key insights and best practices to important stakeholders. Consequently, through the SDGs, the world aims to achieve the remaining targets of the Millennium Development Goals (MDGs) by 2030. The SDGs are recognized as a cocreated strategy of action for global citizens, the world, and the nation's wealth [33]. HEIs are imperative to addressing sustainable development challenges by exceeding training development and skills acquisition and advancement [21]. HEIs must not be restricted to producing exceptional educators; instead, they should also seek to "uncover ground-breaking research and connect services to communities" [18: 418] because HEIs tend to stay in neutral territory in the eyes of various stakeholders and are considered to be among the important hubs for social, economic, and other types of development in any country [34]. In this regard, HEIs are incentivized to formulate management systems grounded in the values communicated in the 17 SDGs.

4 SDG Localization in the Curriculum, Research, Outreach, and Campus

HEIs are currently involved in a multitude of measures to advance sustainability. These measures include embedding sustainability matters into the curriculum, research, and the physical campus and engaging in outreach activities. For example, a survey conducted in 167 universities from various parts of the world on the integration of the Agenda 2030 SDGs with sustainability teaching indicated that lectures are the most prevalent method of integrating the various aspects of sustainability [15]. However, other measures that could directly influence society seem to be less prevalent, such as those tailored to improving the capacities of educators to teach and encourage students to construct sustainable futures and foster an ecosystem for inter- and multidisciplinary research to address highly sophisticated problems [48]. Researchers have also found that measures to attain sustainability in the university setting are mainly centered around operational activities and technological solutions, such as the creation of green spaces and landscapes across campuses. In comparison, efforts aimed at encouraging HEIs to consider behavioral and cultural issues in the organizations themselves are usually nonexistent but considered imperative to a shift toward sustainability [26].

With the primary aim of quickly tracking the localization of the SDGs in HEIs, the chapter combines both conceptual and case study-based insights into the localization of the SDGs. The following sections are devoted to identifying how HEIs can strategize the SDGs within curricula, teaching, and learning, as well as research and development, along with overall SDG localization outside the highlighted dimensions.

4.1 Teaching and Learning Space

Tandon [30] proposed several feasible measures that could be implemented by HEIs to incorporate SDGs into the pedagogy and learning sphere. Tandon [30] recognized the main pathways as the development of curricula, introduction of new courses, and interactive pedagogy. The underlying argument with the development of the curriculum is that the current syllabi and curricula can be further adjusted to improve new perceptions from the SDGs not considered in old curricula. From the perspective of interactive pedagogy, the teaching of numerous SDGs might occur outside the confines of traditional lecture venues and can be practically shown to function within community settings [21]. For example, agricultural science departments in HEIs could engage with their surrounding communities to provide education on organic farming and traditional food conservation practices that conform to the values shared in SDG 2.

Business schools, for example, can facilitate translating the SDGs into practical and workable solutions that resonate with the surrounding communities. Consequently, this goal would require integration of the SDG agenda into the curricula of schools of business. Such incorporation would be vital for a future in which it likely will be possible to share a common worldview that working within the SDGs is a license to function and operate. The SDGs could also be utilized as an avenue for business schools to facilitate external stakeholders, helping them with SDG localization in their own activities [21]. Weybrecht [38] suggested a mechanism for sustainability engagement, which included four steps: (1) setting the scene; (2) integrating (embedding, collaborating, and contributing); (3) identifying unique avenues of engagement; and (4) providing an empowering ecosystem. This mechanism could potentially require important stakeholders, such as governments, corporate leaders, and United Nations agencies, to provide support in shaping future leaders that would help to achieve the aims of the SDGs [19].

4.2 SDG Localization in the Research and Development Space

Research in HEIs has played a considerable role in addressing sustainability through numerous projects and research advancements and via the integration of sustainability principles into practices [41]. HEIs have created research capacities and promoted innovation and interdisciplinary research around the SDGs [5].

Recent work on SDG baselines [17, 20, 42] (Nhamo et al., 2020) has indicated that several countries lack acceptable baselines for measuring, reporting, and verifying (MRV) SDG progress. Additional findings from this work are on significant gaps in data for the indicators devised for MRV that entail big data. Three practical ways in which HEIs can cause research and development to have a more effective impact in understanding SDGs are suggested. These ways encompass: (1) framing locally usable research; (2) coproducing knowledge through partnerships; and (3) learning new competencies [30].

Students and faculties are equally incentivized to formulate policy and highly relevant research questions. Given the unique types of governance systems between nations, research questions related to the SDGs should seek to define and address issues from the global, regional, and national settings. For example, numerous questions could emerge, such as how can research conducted in HEIs addresses issues of extreme weather events arising from climate change, as well as teaching and learning for the SDGs [20].

5 The Evolution of the State of Qatar as an Education Hub

Higher education is imperative in the shift toward a knowledge-based society, and it has held a key position in developing and establishing Qatar as an education hub. The government perceives education as an integral tool for modernization of the nation, diversification of an economy dependent on oil and gas, and cementing the international competitiveness of its citizens. Consequently, the educational sector has been experiencing swift transformation and development through a top-down approach. The effective development and transformation of Qatar's educational sector have been orchestrated by the Qatari leadership through careful and systematic measures. Recognizing the need for more effective action, the Qatari government commissioned RAND in 2001 to examine the educational system and recommend reform solutions to address the system's deficiencies [6]. The objective was to create an education system that would provide young people with the necessary skills to participate more actively in the nation's economic and social systems [6]. RAND's analysis pinpointed the existing system's strengths and flaws and recommended two primary reform priorities: enhancing the education system's core aspects through a standards-based system and developing a system-changing plan to address the system's overall deficiencies [16]. According to the RAND evaluation report, Qatar's education system was restrictive and lacked standards and international benchmarks. Another major weakness identified by the analysis was a lack of vision for the development of the educational sector, which originated from the extremely centralized Egyptian model for education [7]. In addition, there was little consideration given to other models or strategies to construct a high-quality education system that met the needs of stakeholders [7]. Furthermore, development within the education system was piecemeal without considering the whole system as one cohesive unit, and the Ministry of Education suffered from a poor hierarchical organizational structure that did not encourage innovation and reform [7]. The Qatari leadership then transformed the educational system into a new national educational system based on RAND's recommendations to improve the quality of teaching and raise student achievement [16]. Within the Qatari leadership, Sheikha Mozah Bint Nasser Al Missned has been instrumental in the development of Qatar's education sector [6]. The Qatar Foundation was established in 1995 with Sheikha Mozah as the head of its board of directors [6]. The Qatar Foundation is a nonprofit organization that houses and guides various centers and programs centered around education, innovation, and research, with the key aim of unlocking human potential and fostering social development [22]. More recently, the Qatar National Vision (QNV) 2030 has played a pivotal role in shaping the nation's educational sector. Launched in 2008, QNV 2030 positions education as a key pillar for social development in Qatar. The Ministry of Education and Higher Education intends to continue to provide financial and human resource support for educational development as part of its long-term strategy [23]. Consequently, Qatar has initiated efforts in the following aspects [23]:

- Creating a world-class educational system that adheres to contemporary international standards and provides opportunity for citizens to increase their capacities;
- Increasing the numbers of educational facilities and research opportunities;
- Constructing new early childhood centers, schools, colleges, and institutions to help citizens and residents to attain a higher level of education;
- Increasing its education budget allocation to more than 13% of gross domestic product (GDP); and
- Increasing interest in educating students with disabilities, primarily through the designation of additional schools as integration centers.

Since its independence, Qatar has established itself as a hub for skilled labor. The primary objective of establishing an education hub is to shape and promote the country's shift to a knowledge-based economy through training and strengthening the skills and capabilities of the highly educated workforce. The education hub plays a vital role in enhancing the attainment of higher education among Qatari citizens and ensuring that there is no shortfall in skilled human resources for the country's workforce and, more importantly, its economy. Qatar's journey toward cementing its position as an education hub has transformed it into a nation that houses more than ten foreign academic institutions, including branches of Texas A&M University, Carnegie Mellon University, Weill Cornell Medical College, Northwestern University, Georgetown University School of Foreign Service, HEC Paris, Virginia Commonwealth University School of the Arts, the University of Calgary, the University of Aberdeen, and Stenden University. In addition, Qatar is home to top science and research institutes and at least 8000 students and researchers from the Gulf, as well as all around the globe. Qatar, as an education hub, holds numerous achievements and portrays great potential for the future, with its proactive approach toward developing various research and industrial organizations, such as the QSTP and the SMRC, and its numerous initiatives to combine higher education, research, and commercial activities. However, the question persists regarding whether a set of educational reforms could be effective tools for the desired aim of economic and social development [46].

Until 2001, Qatar University was the only existing educational institution in the country for higher education. Over the last decade, foreign universities have also opened campuses in Qatar, thereby increasing the range of options for higher education for Qatari citizens.

The education hub project is not only a representation of the small Gulf country's ability to establish foreign campuses of prominent universities since it also systematically integrates academic programs that feature Education City, research initiatives funded by the Qatar National Research Fund, and research institutions, such as the Qatar Foundation's Qatar Science and Technology Park and the Sidra Medical and Research Center [46].

5.1 Qatar University as a Case Study

The State of Qatar, as part of the greater international community, is also moving steadily toward incorporating sustainability into its strategic plans and activities, including those related to higher education. Education is the driving force establishing sustainability since it is one of the main communication vehicles and the basis for the "sustain-ability mindset" [41]. QU was founded almost four decades ago as a college of education, with a founding cohort of 57 male and 93 female students (Qatar University, 2021). It has since grown to include 10 colleges encompassing fields ranging from engineering to health sciences and Islamic studies. Since its inception, QU has been transformed into a multidisciplinary university that covers numerous academic fields, with the most recent discipline being dental medicine (established in 2020). The education system of QU is closely aligned with the North American higher education system, and most faculty members and researchers are expatriates. QU started as a bilingual institution (Arabic and English) and moved to the English language as a medium of instruction in 2003. Further, in 2012, it adopted the Emirati decree to Arabize a number of arts, business, and law programs. QU implemented significant measures to enhance the quality of education, establish global academic practices, and shift in the direction of a university that is particularly tailored to research. Numerous fundamental changes have been put into practice as part of the reform. These changes include formulating a core curriculum, standardizing testing requirements for admission, merging colleges, and forming a faculty senate [46]. Today, QU is home to more than 22,000 students.

Since QU plays a crucial role as a major hub for national human and knowledge development, it plays an essential role in sustainability in Qatar, making it an interesting case study. Being rooted at the local level and connected at the global level, QU is considered a key agent in the education of future leaders and generations in the implementation of the SDGs. QU has been working in recent years toward the integration of sustainability into its processes and practices. Such efforts have varied from a green campus perspective and the involvement of the university community to transforming academic programs and institutional policies fostering a sustainable post-carbon agenda since the university perceives itself as playing a leading role in the transition from an oil-based economy to a knowledge-based economy.

Furthermore, QU is aiming to play a leading role in the transition toward sustainability to the benefit of society. QU leadership is aware of the world being increasingly characterized by global change impacts, such as income inequality, modern risks, and climate, health, and economic risks. It is therefore crucial for QU to consider the role of academic institutions in having graduates who meet the requirements for more people-centered development in a changing world, in which people's capabilities and voluntary initiatives can decide the outcomes of key global challenges. Such consideration by QU has translated into the above-explained implemented strategies to take teaching and learning to a different level with special focuses on providing significant opportunities for professional development of academic staff/faculty in relation to using educational technologies, teaching, learning, and assessment of learning outcomes. QU has significantly invested in its physical campus and the construction and design of state-of-the-art facilities, in addition to training faculty to be qualified to teach students the critical twenty-first-century skills that learners require to live and work in a globalized society. The following sections aim to highlight the role played by QU and its practices in relation to incorporating sustainability.

5.2 Strategies and Institutional Policy Perspectives on the SDGs

QU aligned its mission and strategic plans to the Qatar National Vision (QNV) of 2030 and its roadmap (Qatar National Development Strategy (QNDS)), both of which consider sustainability among their foremost objectives. The state, through QNV, has a clear goal of preserving the environment and its natural balance to accomplish comprehensive and sustainable development for all generations.

QU has worked extensively in recent years to demonstrate a greater commitment to promoting sustainability, including teaching and learning, academic programs, research, institutional policies, strategies, operations, reporting, and collaboration. The 2018–2022 QU transformation strategy was created considering the digital era and the fourth industrial revolution as external drivers, and its vision has been extended to consider QU as a catalyst to achieve sustainable socioeconomic development in Qatar. For example, QU is fostering effective engagement with local and international stakeholders to enrich education, strengthen research, and impact socioeconomic development and to excel in research that is focused, relevant, measurable, solution oriented, and impactful and that fosters interdisciplinary collaboration and advances knowledge and innovation for the benefit of society. The university is implementing holistic education that is transformative based on implementing six excellence themes: learner-centric, experiential, research-informed, competency-based, digitally enriched, and entrepreneurial (QU Strategic Plan 2018–2022, p. 20). As part of QU Strategy 2018–2022, the University has approved a new QU Qualifications and Competences Framework. This framework outlines the competencies, knowledge, and skills to be acquired by QU students to better acquire twenty-first-century skills that will enable them to maximize their future impact locally and internationally and to prepare them to lead their country's development. These attributes include competence, entrepreneurs, lifelong learning, roundedness, and ethical and social responsibility. All QU programs are expected to demonstrate their alignment with this framework or undertake the necessary actions to ensure that these attributes and underlying supporting competencies are achieved in their programs. Furthermore, QU is to be commended for its commitment as the only national university in the country educating Qataris, especially women, to fulfill the aspirations of its citizenry and the nation, particularly in educating women in all fields.

HEIs are usually criticized in their agendas for sustainability since they approach it in isolated silos [44]. QU is an interesting case study since it has worked on the purpose of integrating sustainability into teaching and learning, as well as in research, holistically. The leadership of QU recognizes that, by conceptualizing the university as a socially responsible institution, any institutional, human capacity, or knowledge deficit can be addressed. Therefore, QU realizes that such incorporation of HEIs into the localization of the SDGs from a social responsibility perspective should be communicated via the three comandates of any HEI: teaching and learning, research, and engaged stakeholders (also known as community engagement or service to the community). Nevertheless, the interaction among these mandates is highly sophisticated in both the conceptual and practical senses (Nhamo, 2012). Without considering the overlapping features of human development, education, sustainable development, and education for sustainable development, an emphasis on SDG localization through HEIs will be highly ineffective [20]. The lifelong learning dimension from SDG 4, which is seen as instrumental in reaching the targets of many other SDGs, can be added [10]. Along with SDG 4, SDGs 3, 5, 12, and 13 have strong connections to lifelong learning.

Apart from the broader sustainability agenda, QU recognizes that there is an explicit SDG (SDG 4) targeting education. SDG 4 implies ten targets and several indicators, and some of the targets include the following [33: 17]:

- Equal access to affordable and high-quality technical, vocational, and postsecondary education for all women and men;
- Increasing the proportions of youths and adults with relevant skills, including technical and vocational skills necessary for employment, decent work, and entrepreneurship;
- Eliminating gender gaps in education and ensuring that disadvantaged groups, such as persons with disabilities, indigenous people, and children in vulnerable situations, have equitable access to all levels of education and vocational training; and
- Globally, increasing the number of scholarships accessible to people from developing nations for higher education enrollment.

To achieve the above targets and, to a certain extent, the broader SDGs, QU has adopted five strategies to facilitate the localization of SDGs. Such strategies include the improvement of HEI quality, equity, environment, research and innovation, and partnerships at the local and global levels, as detailed in the following sections.

5.3 Localizing SDGs in Teaching and Learning

At QU, sustainability education is widely included and embedded in program curricula. In terms of programs, QU has established in recent years academic programs and courses related to sustainability that aim to promote sustainable practices. The bachelor's degree program in environmental sciences, which is the first program of its kind in Qatar, has witnessed a considerable increase in the number of registered students over past years by 20%. In addition to a number of postgraduate programs, such as master of science programs in environmental science and materials science and technology, as well as PhD programs in biological and environmental sciences, and environmental engineering, a significant number of sustainability-related courses have been added to the general education component of the curriculum to raise awareness among QU students about the economic, political, cultural, and social issues impeding the development of a more sustainable future [24]. These engagements represent both a challenge and an opportunity for faculty members, who must incorporate sustainability into their teaching, as well as research. Courses in biological sciences addressing sustainability have increased their students' capacity by 26%, while courses in engineering related to sustainability have undergone an increase of 17% since 2016. More interestingly and in line with the literature, the number of students in the College of Business and Economics registered in courses related to sustainability has increased by 40% as of 2022. Table 1 highlights the initiatives undertaken by QU in engaging the SDGs in teaching and learning.

Furthermore, the growing demand from employers and governments for a more skilled workforce has greatly influenced the role of higher education in society. As a result, HEIs are asked to demonstrate that they are providing capable, competent, and informed citizens able to address the challenges of contemporary life [8, 9, 14]. To demonstrate their suitability, QU adopted an outcome-based approach to education, graduate attributes, and competency frameworks with the aim of preparing students for twenty-first-century skills and adapting to future uncertainties. This strategic change has considerably helped QU to work on achieving SDG 4 by offering to its students a quality education that will help them to be ready to adjust to the constantly changing workplace. QU has also developed a white paper about labor

Table 1 Initiatives by Qatar University in engaging the SDGs in teaching and learning, research, and community service

Domain	SDGs' initiative
Curricula and teaching and learning	Teaching SDG 13 in a standalone course entitled "Climate Change Policy and Analysis"
	Teaching SDGs 1, 7, 11, 12, 14, and 15 as modules in the Policy, Planning, and Development Program
	Initiating a proposal for a multidisciplinary postgraduate degree in "Sustainable Development"
	Mapping some of the programs to the SDGs.
	Implementing a Teaching and Learning Strategy university-wide, including graduate attributes and supporting competencies, such as subject matter, adaptive thinking, problem-solving, creativity, critical thinking, collaboration, self-awareness, and ethical and social responsibility. These skills should prepare students to solve twenty-first-century global challenges (QU Qualification Framework, 2019) and eventually to become effective future SDG implementers.
	Organizing conferences and a series of public lectures on the SDGs to promote and support the principles of the SDGs, to call for sustainability in the regional and global context, and to create an awareness about the significance of their implementation

market alignment in Qatar to study the need for workplace alignment with transforming Qatar into a "knowledge-based economy" based on Qatar National Vision 2030. The white paper provides a descriptive analysis and correlates data extracted from different official reports at a national level with published reports from QU. Using this holistic approach to situate QU within the labor market helps in initiating the project, aligning the number of graduates with job vacancies, and matching/enhancing the program's curriculum with the skills needed and meeting the future national demand for workers. A performance metric/model for labor market alignment efforts was among the major outcomes of the project.

5.4 Localizing SDGs in Research

The implementation of sustainability has also occurred through research projects encouraging undergraduate and graduate students, as well as faculty members, to become involved in research that promotes sustainability. QU has funded several research projects and organized several workshops and conferences related to sustainability with collaborations and the participation of local and international entities. These events include the international sustainability workshop in October 2021; the design e-workshop in May 2020 and 2021; and the Food, Energy, Water, and Waste Workshop in April 2021; in addition to participation in the 56th ISOCARP World Planning Congress: PostOil City: Planning for Urban Green Deals in November 2020; and the Sustainable FEW Nexus and Food Waste Management in an Urban Context Workshop in February 2020. QU hosted the Science for Sustainable Development 2019 Conference (SSD2019) in November 2019, which was oriented toward integrated and sustainable resource management. Students and all QU communities were encouraged to participate and be involved in activities or events oriented toward understanding and facilitating sustainability in Qatar and worldwide. These events included the "University Campus Grows Green" initiative; the organization of the "Intelligent System for Solar Energy (Green Technology)" competition, which targeted high school students; participation in Qatar National Environmental Day 2021, which involved different activities; and two competitions and a campaign to plant 1 million trees in the country. Given such commitments, students and faculty members were invited to engage more actively with sustainability and these initiatives. Table 2 lists the initiatives by QU in engaging the SDGs in research.

5.5 Localizing SDGs in Institutional Policies, Strategies, and Operations

QU has worked on reconsidering its governance and policies to adopt a more sustainable and modern approach to respond to and anticipate the challenges of a complex and rapidly changing world. To address SDG 5 (gender equality), it has

Table 2 Initiatives by Qatar University in engaging the SDGs in research

Domain	SDGs initiative
Research	Investing millions of dollars in leading studies in sustainable development research, producing SDG-related knowledge, technology, and innovation plans specifically for SDGs 3, 4, 7, 11, 13, and 14
	Mapping research to SDGs since QU has adjusted its research priorities to make them better aligned with the national research priorities focusing on the transition to sustainability in all domains and addressing the global challenges facing this transition
	The Center for Law and Development at QU is undertaking research in energy and environment that provides solutions to sustainable development challenges aligned with SDGs 7 and 13. https://www.qu.edu.qa/research/cld/about/development/energyandenvironment
	The Center for Sustainable Development at QU is undertaking specific projects and services funded by the Qatar National Research Fund and from within Qatar University, as well as industry-funded projects with companies such as Exxon Mobil, Total, QDVC, and Marubeni. These projects are aligned to the global sustainability agenda, particularly to SDGs 7 and 13, mainly focusing on algae biofuel technologies, climate change mitigation and adaptation, a sustainable food-energy-water nexus, food waste management in an urban context, and natural resource governance.

enforced nondiscrimination and equality between women and men at the levels of faculty members, students, and staff in the recent issue of its faculty bylaws, professional code of conduct bylaws, student code of conduct, and the business operations department policies handbook [25]. QU focuses on educating women in all fields, not only to fulfill the development needs of the country but also to achieve gender equality in education. Furthermore, QU's scholarly output in SCOPUS increased to 56 in 2020 regarding gender equality, and the rate of international collaboration concerning this SDG reached 53.6%.

Reduced inequalities have also been the subject of several publications by faculty members at QU, where the scholarly output in SCOPUS from 2016 to 2020 grew to 137. QU has also paid particular attention to people with special needs in terms of employment opportunities, student facilities, support, and equality between all staff in rights or privileges in workplaces, as well as facilitating different procedures or services. Measures against discrimination have been addressed in a variety of legal institution-wide documents in QU [25]. Another important unit at QU, which is the inclusion and special needs support center, has an ultimate goal of ensuring education and services for all.

QU has also paid particular attention to SDGs 7 and 12. QU has reviewed its regulations related to waste collection disposal, as well as hazardous waste disposal and sanitation. The health, safety, and environment office has set a new mission to provide the community of QU with a sustainable environment by maintaining the highest possible standards. It recently issued a policy on renewable energy and conservation to set regulations to provide convenient educational and research environments while maximizing efficiency and responsibly managing energy, water, and human resources. Moreover, QU has a number of waste prevention efforts, mainly under the Zero-Waste Initiative, governed by a zero-waste policy. From research publications, the scholarly output in SCOPUS has dramatically increased to an average of 1200 for both SDGs, with a view count that reached more than 45,000, as well as international collaboration in the area of clean energy that reached 72.7%. Table 3 lists some initiatives undertaken by QU in engaging the SDGs in collaboration and outreach.

QU is constantly ranked in the top 300 and entered the top 200 in 2021 of the Times Higher Education (THE) University Impact Ranking, which was first introduced (and piloted) in 2019. The university is using this ranking to reframe its considerable research portfolio related to the SDGs and to help categorize the research by SDG target [35, 36]. It is important to note that the Times Higher Education's Impact Rankings measure a university's contribution to society in terms of its achievement in achieving the UN SDGs. Such an initiative by QU provides a showcase and an opportunity to shine a light on institutional activities and efforts not captured by other rankings. The ranking in itself helps the university to benchmark its efforts on specific SDGs against the best universities in the world achieving considerable results, especially in SDGs 2 (84th), 6 (51st), 11 (96th) 14 (53rd), 15 (54th), and 17 (52nd). The obvious importance of this ranking in terms of providing a tool to universities for reporting on the SDGs is, however, nuanced. Its methodology is based on a subjective evaluation of the documents submitted and is a subject of yearly changes, limiting its reliability.

The limitations of this ranking and the need for university administrators to have a more objective tool to transform their research reporting to an SDG-oriented tool caused QU to introduce a platform to "map" and reflect how the university contributes to the implementation of the SDGs in its research [35], educational, and policies and operations. Such a platform will reflect all of the activities, practices, and outcomes by faculty members, staff, and administrators aligned with the SDGs. This mapping will be a very valuable exercise in planning for ongoing implementation of the SDGs and the university's role in this journey. The platform will also identify key persons and departments that, more than others, are engaged in the localization of the SDGs; it will also serve as a mechanism for categorizing key stakeholders and creating a database of the activities, actions, and polices undertaken on campus. Furthermore, it will assist the university administration in determining areas of common importance and interest across the university and prospects for internal partnerships and external collaborations. The university plans to use this platform as a diagnostic tool to define the assets and strengths, as well as gaps, of QU performance to build a measurable process for communicating, reporting, and showcasing QU's engagement and its impact on the SDGs. It also seeks to identify opportunities for future partnerships, collaborations, and actions.

Table 3 Initiatives by Qatar University in engaging the SDGs in collaboration and outreach

SDG 1. No Poverty			
SESRI contribution to the Local Ministry Policy	SESRI is a significant contributor to Qatar National Vision	Guest Workers' Welfare Index Households	Qatar University provides financial assistance to local sustainable businesses
SDG 2. Zero Hunger			
QU Center of Sustainability Development at World Food Day	QU to develop Qatar's first solar agriculture greenhouse	College of Engineering hosts seminar on "Food Security, Technology and Techniques"	COVID-19 Symposium: Supply Chain Disrupted and Food Security
SDG 3. Good Health and Wellbeing			
Sixteen QU alumni volunteered at National Volunteering Campaign COVID-19	QU and Sidra Medicine sign MoU training for the next generation of health care professionals	QU conducts first national mental health study	QU Health Clinic organized an event for its students and faculty dedicated to mental health
SDG 4. Quality Education			
Qatar University's Key Collaborator Virtual Education Forum 2021	QU's Continuing Education Center offers online courses for free	QICCA, Qatar University's CCE, signs a deal to implement Arabic training programs	Qatar University and ESCWA host first meeting of Academic Network for Development Dialogue
SDG 5. Gender Equality			
QU hosts UNESCO Women and Girls in Science in Qatar	QU is developing a platform to promote entrepreneurship for women in Qatar	QU Women's Entrepreneurship Day	Symposium on Efforts to Combat Discrimination against Women
SDG 6. Clean Water and Sanitation			
QU organizes workshop on the Food-Energy-Waste-Water Nexus	QU marks World Water Day	Qatar University and ExxonMobil develop industrial wastewater treatment using algae	Qatar University faculty members highlight importance of water conservation
SDG 7. Affordable and Clean Energy			
QU and Ministry of Environment work on biofuels	New patent to remove carbon monoxide from car exhaust	TOTAL and QU launch Research on Carbon Capture & Biofuels	QU Call for Papers: Targeting SDG7 Policy Development Goals
SDG 8. Decent Work and Economic Growth			
SESRI & ILO: The Wage Protection System in Qatar	QU announces results of the guest Workers' Welfare Index 2018	The SDGs and the misuse of artificial Intelligence	QU hosts the 6th Annual International Conference of the Gulf Studies Center

(continued)

Table 3 (continued)

SDG 9. Industry, Innovation and Infrastructure			
QU participates in St. Petersburg International Economic Forum	QU hosts 9th Annual Innovation and Entrepreneurship Contest	"I'm Research" Program	QU honors more than 50 distinguished QU scientists
SDG 10. Reduced Inequalities			
Distance learning for special needs students	QU marks World Autism Day	QU hosts the Future of Inclusive Education in Arab Countries	QU's "Yes, I Can" Campaign
SDG 11. Sustainable Cities and Communities			
QU to design and implement innovative cooling systems for FIFA World Cup stadiums	UAV-enabled Intelligent Transportation Systems for the Qatar Smart City	QU and Museum of Islamic Art (MIA) on cultural heritage in the digital era	QU Arts and Heritage: QU forms a committee to collect Qatari Heritage from the 1950s and 1960s
SDG 12. Responsible Consumption and Production			
QU initiative produces fertilizer using food waste from campus	QU designed and constructed a vegetable factory pilot plant	QU's Wastewater Management Technology	QU wins award at Fourth Qatar Sustainability Awards 2020
SDG 13. Climate Action			
QU is a participant in the United Nations Framework Convention on Climate Change	Climate change and the GCC	QU's legal clinic conducts study of Qatari climate change laws and policies	QU to develop Qatar's first solar agriculture greenhouse
SDG 14. Life Below Water			
UNESCO chair at QU for Marine Sciences	"QataREEF" Project	Hawksbill Turtle Conservation	Developing coral reef restoration facilities in Qatar
1.8 Million Dollar Fund	"Saving Marine Spices at Risk in the Gulf"	QU and Industrial Partners in Qatar	Developing cutting-edge research to study the whale shark populations
SDG 15. Life on Land			
DUGONG preservation project	QU initiatives on monitoring and reducing oil pollution	QU SESRI implementation of the Agricultural Census in Qatar	QU researchers develop biopesticides
SDG 16. Peace, Justice and Strong Institutions			
QU & UNODC-Education for Justice initiative	Shura Council elections and future vision	QU–SESRI policy contribution to State of Qatar	QU call for papers: A Focus on SDG 16
SDG 17. Partnership for the Goals			
QU's contribution to Qatar's Voluntary National Review 2021	QU launches ANDD with UN (ESCWA)	SDGs at Qatar University	QU and French embassy hold seminar on climate change

6 Conclusion

Through systematic collaboration with faculty, students, alumni, and the surrounding stakeholder community, universities can play a more proactive role in meeting the SDGs. Currently, universities play an important role in paving new pathways for the world, driving a sense of "global citizenship" and fostering and supplying knowledge and innovation to the community. As such, universities have the potential to become major drivers of societal transformation.

This chapter aimed to investigate the various methods of localizing sustainability in achieving the SDGs in the university context with a case study from the Gulf region. One of these methods was to devise a tool enabling administrators and decision-makers to have an overview of the existing research and map it by SDG. It would also allow them to use the collection of data to put in place a road map to transform the research done from reactive to the SDG implementation to a proactive one serving the needs of its society.

The chapter highlighted the transformative role of HEIs in achieving the SDGs by balancing the scientific foundation and technology tools to transform their campuses into smart communities with the long-term strategies adopted in academia and research to establish a sustainability culture that will influence the next generation. In particular, the case study of QU, the first and largest national higher education institution in the country, was used. With the primary aim of quickly tracking the localization of the SDGs in HEIs, the chapter combined both conceptual and case study-based insights into the localization of SDGs. The chapter explored the different strategies implemented by QU and how its own assets and constituencies within its campus, including its operational and academic arenas, were used to overcome sustainable development challenges. The significant investment in infrastructures, the construction and design of state-of-the-art facilities, and the incorporation of sustainability into its strategic plans and activities have significantly established a culture of sustainability within the campus. The involvement of the university community in transforming academic programs and institutional policies fostering a sustainable post-carbon agenda has promoted the national role of the university in the transition from an oil-based economy to a knowledge-based economy.

These strategies have demonstrated the role played by QU and its practices in relation to incorporating sustainability, with the aim of having graduates who will meet the requirements for more people-centered development in a changing world where people's capabilities and voluntary initiatives can decide the outcomes of key global challenges.

QU was an interesting case study since it has worked on the purpose of integrating sustainability into teaching and learning, as well as in research, holistically. The leadership of QU recognizes that, by conceptualizing the university as a socially responsible institution, any institutional, human capacity, or knowledge deficit can be addressed. Therefore, QU realizes that such incorporation of HEIs into the localization of SDGs from a social responsibility perspective should be communicated via the three comandates of any HEI: teaching and learning, research, and community engagement.

The way in which the sustainability mandate of Agenda 2030 is being interpreted and formulated in both discourse and practice by HEIs at the global and regional levels continues to be a topic lacking adequate research and investigation. Consequently, it is imperative to determine whether there are any overlapping notions among HEIs regarding what sustainability should encompass in higher education discourse and practice at all levels. In addition, it is also essential to pinpoint any combined and highly intercoordinated measures among various stakeholders to understand whether there are any commonalities among HEIs regarding what sustainability should cover in higher education discourse and practice and whether there are any corresponding gaps in knowledge and understanding. Adequate research is needed to address these key issues by comparatively examining the sustainability discourses of a sample of important global and regional HEI networks. It will then be necessary to pinpoint the key action points that HEIs are attempting to address and to explore the primary patterns in and significant obstacles to fostering sustainability within higher education. The findings of the SDG mapping in QU indicate that Quality Education (SDG 4) is by far the most desired goal for all recent initiatives in teaching and research and in the university's own policy-making. Additionally, the targets of SDG 5 (Gender Inequality) and SDG 7 (Affordable and Clean Energy) and SDG 13 (Climate Change) are also significantly emphasized in various initiatives. In contrast, SDG 1 (No Poverty), SDG 2 (No Hunger), and SDG 17 (Partnership for the Goals) have either been overlooked or been assigned less importance. This fact raises questions regarding what measures and reforms must be implemented to improve the university's commitment to global development on the strategic level. As such, it is evident that these aspects should be integrated into the new strategic planning and policy-making of QU if the end goal is to have a far-reaching global impact.

In summary, the shift toward sustainability is a highly sophisticated process involving different stakeholders, interests, and agendas. In an effort toward the realization of the SDGs at QU, a higher degree of transition toward sustainability could potentially allow the university to adopt a more proactive role in overcoming sustainability issues and putting SDGs into practice on the societal level. However, sustainability typically takes a backseat to mainstream academic fields, with academic work in the field typically detached from campus activities and community service. This fact translates into various problems and hindrances for universities in incorporating sustainability into their governance and business models. These issues constitute the key area of concern when extensive research is required to illuminate more ways in which a university could potentially work toward the targets of the SDGs. In particular, a highly versatile approach is required to connect the various groups, departments, and other stakeholders within the university, as well as with the external stakeholders that it helps and interacts with at the local, national, and international levels.

References

1. Abulibdeh, A., & Zaidan, E. (2020, March). Managing the water-energy-food nexus on an integrated geographical scale. *Environmental Development, 33*. Elsevier.
2. Abulibdeh, A., Zaidan, E., & Al-Saidi, M. (2019). Development drivers of water, food, and energy nexus in arid regions – The case of the GCC. *Development in Practice, 29*(5), 582–593. Routledge.
3. Barnett, R. (2004). The purposes of higher education and the changing face of academia. *London Review of Education, 2*(1), 61–73. https://doi.org/10.1080/1474846042000177483
4. Barth, M. (2013). Many roads lead to sustainability: A process-oriented analysis of change in higher education. *International Journal of Sustainability in Higher Education, 14*(2), 160–175. https://doi.org/10.1108/14676371311312879
5. Blanco-Portela, N., Benayas, J., Pertierra, L. R., & Lozano, R. (2017). Towards the integration of sustainability in Higher Education Institutions: A review of drivers of and barriers to organisational change and their comparison against those found of companies. *Journal of Cleaner Production, 166*, 563–578. https://doi.org/10.1016/j.jclepro.2017.07.252
6. Brewer, D. J., Augustine, C. H., Zellman, G. L., Ryan, G., Goldman, C. A., Stasz, C., & Constant, L. (2007a). Qatar and its education system. In *Education for a new era: Design and implementation of K-12 education reform in Qatar* (1st ed., pp. 7–32). RAND Corporation. http://www.jstor.org/stable/10.7249/mg548qatar.11
7. Brewer, D. J., Augustine, C. H., Zellman, G. L., Ryan, G., Goldman, C. A., Stasz, C., & Constant, L. (2007b). Analysis of Qatar's education system. In *Education for a new era: Design and implementation of K-12 education reform in Qatar* (1st ed., pp. 33–46). RAND Corporation. http://www.jstor.org/stable/10.7249/mg548qatar.12
8. Chan, R., Brown, G. T., & Ludlow, L. (2014). What is the purpose of higher education? A comparison of institutional and student perspectives on the goals and purposes of completing a bachelor's degree in the 21st century. In *American Education Research Association (AERA) conference*. http://www.dal.ca/content/dam/dalhousie/pdf/clt/Events/Chan_Brown_Ludlow(2014).pdf
9. Cooper, S. (2017). A collaborative assessment of students' placement learning. *Assessment & Evaluation in Higher Education, 42*(1), 61–76. https://doi.org/10.1080/02602938.2015.1083093
10. English, L. M., & Carlsen, A. (2019). Lifelong learning and the Sustainable Development Goals (SDGs): Probing the implications and the effects. *International Review of Education, 65*(2), 205–211. https://doi.org/10.1007/s11159-019-09773-6
11. Fairweather, J. S., & Blalock, E. (2015). Higher education: The nature of the beast. In J. Huisman, H. de Boer, D. D. Dill, & M. Souto-Otero (Eds.), *The Palgrave international handbook of higher education policy and governance* (pp. 3–19). Palgrave Macmillan UK. https://doi.org/10.1007/978-1-137-45617-5_1
12. Findler, F., Schönherr, N., Lozano, R., Reider, D., & Martinuzzi, A. (2019). The impacts of higher education institutions on sustainable development. *International Journal of Sustainability in Higher Education, 20*(1), 23–38.
13. Greenwood, R., Raynard, M., Kodeih, F., Micelotta, E. R., & Lounsbury, M. (2011). Institutional complexity and organizational responses. *The Academy of Management Annals, 5*(1), 317–371. https://doi.org/10.1080/19416520.2011.590299
14. Grosemans, I., Coertjens, L., & Kyndt, E. (2017). Exploring learning and fit in the transition from higher education to the labour market: A systematic review. *Educational Research Review, 21*, 67–84. https://doi.org/10.1016/j.edurev.2017.03.001
15. Leal Filho, W., Lange Salvia, A., Pretorius, R. W., Londero Brandli, L., Manolas, E., Alves, F., Azeiteiro, U., Rogers, J., Shiel, C., & do Paco, A. (Eds.). (2020). *Universities as living labs for sustainable development* (World sustainability series). Springer.

16. Nazzer, R. (2017). Qatar's educational reform past and future: Challenges in teacher development. *Open Review of Educational Research, 4*(1), 1–19. https://doi.org/10.1080/23265507.2016.1266693
17. Nhamo, G., & Mjimba, V. (2020). The context of SDGs and institutional higher education. In G. Nhamo & V. Mjimba (Eds.), *Sustainable Development Goals and institutions of higher education* (Sustainable development goals series) (pp. 1–12). Springer. https://doi.org/10.1007/978-3-030-26157-3_1
18. Owens, T. L. (2017). Higher education in the Sustainable Development Goals framework. *European Journal of Education, 52*, 414–420. https://doi.org/10.1111/ejed.12237
19. Parkes, C., Buono, A. F., & Howaidy, G. (2017). The principles for responsible management education (PRME): The first decade – What has been achieved? The next decade – Responsible management education's challenge for the Sustainable Development Goals (SDGs). *The International Journal of Management Education, 15*, 61–65. https://doi.org/10.1016/j.ijme.2017.05.003
20. Purcell, W. M., Beer, J., & Southern, R. (2016). Differentiation of English universities: The impact of policy reforms in driving a more diverse higher education landscape. *Perspectives: Policy and Practice in Higher Education, 20*(1), 24–33. https://doi.org/10.1080/13603108.2015.1062059
21. Purcell, W. M., Henriksen, H., & Spengler, J. D. (2019). Universities as the engine of transformational sustainability toward delivering the Sustainable Development Goals "Living labs" for sustainability. *International Journal of Sustainability in Higher Education, 20*(8), 1343–1357. https://doi.org/10.1108/IJSHE-02-2019-0103. Emerald Publishing Limited 1467-6370.
22. Qatar Foundation. (n.d.). *About us*. Retrieved from https://www.qf.org.qa/about
23. Qatar Development Bank. (2019). *Educational sector in Qatar*. Retrieved from www.qdb.qa/en/Documents/Education_Sector_in_Qatar_EN.pdf
24. Qatar University. (2022a). *Catalog*. http://www.qu.edu.qa/static_file/qu/students/documents/student-catalog-2021-2022-EN.pdf
25. Qatar University. (2022b). *QU policy/Bylaws portal*. http://www.qu.edu.qa/about/policy/hot-topics
26. Ruiz-Mallen, I., & Herras, M. (2020). What sustainability? Higher education institutions' pathways to reach the agenda 2030 goals. *Sustainability, 2020*(12), 1290. https://doi.org/10.3390/su12041290
27. Shields, R., & Watermeyer, R. (2018). Competing institutional logics in universities in the United Kingdom: Schism in the church of reason. *Studies in Higher Education*, 1–15. https://doi.org/10.1080/03075079.2018.1504910
28. Sin, C., & Amaral, A. (2017). Academics' and employers' perceptions about responsibilities for employability and their initiatives towards its development. *Higher Education, 73*(1), 97–111. https://doi.org/10.1007/s10734-016-0007-y
29. Sin, C., Tavares, O., & Amaral, A. (2017). Accepting employability as a purpose of higher education? Academics' perceptions and practices. *Studies in Higher Education*, 1–12. https://doi.org/10.1080/03075079.2017.1402174
30. Tandon, R. (2017). *Making the commitment: The contributions of higher education to SDGs*. UNESCO.
31. Trencher, G., Yarime, M., McCormick, K., Doll, C., & Kraines, S. (2014). Beyond the third mission: Exploring the emerging university function of co-creation for sustainability. *Science and Public Policy, 41*(2), 151–179.
32. UNESCO. (2017). *Education for Sustainable Development Goals: Learning objectives*. UNESCO.
33. United Nations. (2015). *Transforming our world: The 2030 agenda for sustainable development*. United Nations Secretariat.
34. Utama, Y. J., Ambariyanto, A., Zainuri, M., Darsono, D., Setyono, B., Widowati, S., & Putro, S. P. (2018). Sustainable development goals as the basis of university management towards

global competitiveness. Journal of Physics: Conference Series, 1025(1), 012094. https://doi.org/10.1088/1742-6596/1025/1/012094
35. Wazen, C. (2020a). *Did the overall ranking mess it up?* Retrieved from https://www.linkedin.com/pulse/did-overall-ranking-mess-up-cesar-wazen
36. Wazen, C. (2020b). *Did the overall ranking mess it up? Addendum.* Retrieved from https://www.linkedin.com/pulse/did-overall-ranking-mess-up-addendum-cesar-wazen
37. Weiss, M., & Barth, M. (2019). Global research landscape of sustainability curricula implementation in higher education. *International Journal of Sustainability in Higher Education, 20*(4), 570–589. https://doi.org/10.1108/IJSHE-10-2018-0190
38. Weybrecht, G. (2017). From challenge to opportunity management education's crucial role in sustainability and the Sustainable Development Goals – An overview and framework. *The International Journal of Management Education, 15*, 84–92. https://doi.org/10.1016/j.ijme.2017.02.008
39. Zaidan, E., & Abulibdeh, A. (2020). Master planning and the evolving urban model in the Gulf cities: Principles, policies, and practices for the transition to sustainable urbanism. *Planning Practice and Research.* https://doi.org/10.1080/02697459.2020.1829278
40. Zaidan, E. (2019). Cultural-based challenges of the westernized approach to development in newly developed societies. *Development in Practice, 29*(5), 670–681. Routledge.
41. Žalėnienė, I., & Pereira, P. (2021). Higher education for sustainability: A global perspective. *Geography and Sustainability, 2*(2), 99–106. https://doi.org/10.1016/j.geosus.2021.05.001
42. Zhou, L., Rudhumbu, Shumba, J., & Zhou, A. (2020). Role of higher education institutions in the implementation of Sustainable Development Goals. In G. Nhamo & V. Mjimba (Eds.), *Sustainable Development Goals and institutions of higher education* (Sustainable Development Goals series) (pp. 88–96). Springer. https://doi.org/10.1007/978-3-030-26157-3_1
43. Davim, J. P., & Leal Filho, W. (Eds.). (2016). Challenges in higher education for sustainability. Springer.
44. Dmochowski, J. E., Garofalo, D., Fisher, S., Greene, A., & Gambogi, D. (2016). Integrating sustainability across the university curriculum. International Journal of Sustainability in Higher Education, 17(5), 652–670. https://doi.org/10.1108/IJSHE-10-2014-0154
45. Fumasoli, T., & Stensaker, B. (2013). Organizational Studies in Higher Education: A Reflection on Historical Themes and Prospective Trends. Higher Education Policy, 26(4), 479–496. https://doi.org/10.1057/hep.2013.25
46. Ibnouf, A., Dou, L., & Knight, J. (2014). The Evolution of Qatar as an Education Hub: Moving to a Knowledge-Based Economy. In J. Knight (Ed.), International Education Hubs: Student, Talent, Knowledge-Innovation Models (pp. 43–61). Springer Netherlands. https://doi.org/10.1007/978-94-007-7025-6_4
47. Kezar, A. J. (2001). Understanding and Facilitating Organizational Change in the 21st Century: Recent Research and Conceptualizations. ASHE-ERIC Higher Education Report, Volume 28, Number 4. Jossey-Bass Higher and Adult Education Series.
48. Lozano, R., Lozano, F. J., Mulder, K., Huisingh, D., & Waas, T. (2013). Advancing Higher Education for Sustainable Development: International insights and critical reflections. Environmental Management for Sustainable Universities (EMSU) 2010, 48, 3–9. https://doi.org/10.1016/j.jclepro.2013.03.034
49. Musselin, C. (2007). Are universities specific organisations. Towards a Multiversity, 63–84.
50. Nhamo, L., Ndlela, B., Mpandeli, S., & Mabhaudhi, T. (2020). The Water-Energy-Food Nexus as an Adaptation Strategy for Achieving Sustainable Livelihoods at a Local Level. Sustainability, 12(20). https://doi.org/10.3390/su12208582
51. Parsons, T. (1951). The social system. United Nations Sustainable Development Summit 2015: Sustainable Development Knowledge Platform. (n.d.). Retrieved December 5, 2022, from https://sustainabledevelopment.un.org/post2015/summit

Leveraging Disruptive Technologies and Systems Thinking Approach at Higher Education Institutions

Mhlambululi Mafu

1 Introduction

Predictive learning analytics (PLA) is the statistical analysis of historical and current learner data using intelligent models [1]. This allows the discovery of information and social connections, better understanding of learners, optimizing their learning experience and environment, and predicting and advising learning. This is achieved through statistical techniques like data mining, modeling, and machine learning. According to Picciano (2012), analytics is the scientific process of investigating data to draw conclusions, organizational efficiency, and decision-making [2]. Data sources include Learning Management Systems (LMS) like Blackboard, Canvas, EPIC, Moodle, Brightspace, Sakai, and Schoology [3, 4]. Through the LMS, students can study from wherever they are and at their own pace, creating an immersive experience through blending traditional and non-traditional teaching methods, which is promising to be the norm. However, the conventional teaching methods do not match novel technology and e-learning environments as future universities move to student-focused learning. Notably, students can study via multiple modes to suit their lifestyles [3]. The systems thinking approach examines complex problems, and systems thinking focuses on interactions among system components and analyses patterns emerging from such interactions [4]. Thus, it exploits data to derive a useful construct, a system, that creates efficiency and utility. Therefore, systems thinking can assist students in developing higher-order thinking skills needed to comprehend and resolve complex challenges as they take charge of their learning.

HEIs can develop courses or learning support processes using the systems thinking approach and PLA to manage complex scenarios and identify hidden patterns and trends from historical data [5]. This allows HEIs to make predictions with

---

M. Mafu (✉)
Department of Physics, Case Western Reserve University, Cleveland, OH, USA

reasonable accuracy about how learners are likely to perform in future academic programs and possible future jobs. Also, PLA allows examination of learner's data and establishes intervention processes to improve student performance, retention, and success [6, 7]. Besides conventional reporting learner's performance, it predicts everything from future student learning outcomes to whether a learner will continue in each program and eventually obtain a degree [8, 9]. Consequently, PLA offers new ways to empower students, faculty, and administrators with intelligence to achieve more desired outcomes. These include evaluating students' learning services, success rates, and support requirements to reduce dropout and failure rates [10]. Thus, universities must provide complete and rounded education so that students acquire the knowledge and competencies required when they graduate and be ready for the nuances of the workplace. This demonstrates that universities' roles have become broader than just knowledge delivery. Essentially, this is what future universities seek to achieve [11].

PLA has not complemented systems thinking though its conceptual framework could potentially be applied in learning analytics, strongly emphasizing understanding systems as a whole and full complexity [12]. Therefore, complementing PLA with systems thinking approach could empower instructors with new sophisticated tools to study teaching and learning, improve course delivery and learning outcomes, and establish student needs, preferences, and behavior. Essentially, this forms student-centered learning. Based on system thinking, at HEIs, there is a complex interaction of various system components, such as formal and informal institutions, processes, people, and social norms and behaviors, and it requires technical solutions. Therefore, we motivate the need to complement PLA with systems thinking approach and support why and how this approach would benefit HEI and stakeholders. However, apart from these benefits, several ethical concerns arise from the analysis, usage, and privacy of data collected from various learning platforms [13, 14]. Other critical issues include sharing and exploiting educational datasets within and outside HEIs for research, technology transfer, and commercialization purposes. This raises an essential need to make learner data more available without jeopardizing their privacy and independent power over their learning [15, 16].

Based on a landscape analysis of PLA and the systems thinking approach, we examine interactions between learners, course design, and future HEIs. We evaluate their potential in improving learning outcomes, supporting learning and teaching, and the degree of their implementation at HEIs. Moreover, we examine how PLA can address HEI concerns during pedagogic provision activities. Furthermore, we investigate how institutional policy development can support a complete implementation of PLA by focusing on the learner, courses and learning design, and future HEI through the systems thinking approach. We examine what PLA and systems thinking can reveal about learning resources, interactions, and learners' participation and assess whether their activities align with expectations derived from pedagogical design principles. Lastly, we focus on the institution and investigate conditions and indicators signaling their learning analytics "readiness" to provide customized, on-demand student-centered education. Therefore, our primary goal is to examine how PLA and systems thinking could assist decide how and when to plan courses and structure programs that meet industry needs. This will allow future universities to produce industry-ready graduates ready to compete for future jobs or collaborate with industries to solve real-world problems. Therefore, except for this introduction, Sect. 2 provides the literature review on how HEIs have leveraged PLA and the systems thinking approach for their competitive advantage. Section 3 discusses the research methodology, and Sect. 4 presents the application and impact of the systems thinking approach and PLA on HEI stakeholders. Section 5 discusses challenges, opportunities, and limitations. Finally, Sect. 6 is the conclusion.

2 Literature Review

Numerous organizations, including HEIs, have begun taking advantage of data that generate potent insights for informed decision-making and competitive advantage [17]. HEI systematically collects and analyses large student datasets to predict how they could tailor their services and personalize learning to improve student outcomes [18]. Walczak (1994) proposed a neural network system, ADMIT, to determine the chance an applicant will attend a specific university if accepted, enabling admissions staff to use time effectively [19]. Integer programming can guide admission staff in making effective decisions [20]. Besides relying on test scores, Fong, Si and Biuk-Aghai (2009) proposed a hybrid model of neural network and decision tree classifier to predict student admission likelihood based on their academic merits, background, and university admission criteria from those historical records [21]. This model could predict suitable universities matching student profiles and providing appropriate entry channels. The model guides and simplifies the complex decision-making process for students and university enrollment officers. Notably, using data to identify and support pain points in the student journey improves student and faculty experience. It revolutionizes the way universities operate, thus creating sustainable future universities.

Furthermore, Ragab, Mashat and Khedra (2012) presented a college admission system using data mining methods and knowledge discovery rules for addressing college admission prediction problems [22]. Huang and Fang (2013) compared four predictive models to predict student academic performance [23]. Using a multiple linear regression model, with students' CGPA as an independent variable. Using an SVM model increased the percentage prediction accuracy and allowed the instructor's interventions to improve student learning. Thus, the role of the instructor must change to more of a facilitator or managing the learning environment. This is a critical cultural shift and is key to achieving the vision for the future. Other works further indicate that PLA can assist instructors and HEIs in overcoming challenges by identifying groups of students requiring additional support to achieving learning outcomes [24]. Scheffel et al. (2017) discovered that visualizing students' performance using the PLA reinforced the learning processes through time [25]. In addition, PLA was used to support instructors in identifying participation problems, facilitating regular interventions, and complementing teaching practice. However,

Conijn et al. (2017) discovered that PLA results differed substantially and established that virtual learning environments (VLEs) engagement and clicking data provide limited value in predicting learning outcomes [26]. This leads to conversations about how they could improve them to serve the stakeholders better.

Using educational data mining techniques, Hassan et al. (2019) and Shahiri and Husain (2015) identified factors influencing student performance using data collected from the LMS and e-learning system [27, 28]. This resulted in the development of a new Student Prediction Model. More efficient student performance and social behavior predictor tool to improve the quality of education in HEIs were developed using data mining algorithms [28]. Based on decision tree analysis, Hamoud et al. (2018) developed an algorithm to investigate factors affecting student success and ultimately predict student performance. To address the high dropout rate in e-learning programs, Tan and Shao (2015) developed a machine-learning method that uses Artificial Neural Network (ANN), Decision Trees, and Bayesian Networks [29]. Their results suggest that all three machine-learning methods effectively predict student dropout, but the Decision Tree presented a better performance. According to Elçi and Abubakar (2021), task-technology fit (TTF) and technology-induced motivation constitute necessary conditions for high student learning performance [30]. This demonstrates the power of advanced analytics in unlocking deeper insights that traditional universities could achieve through descriptive and diagnostic analytics, which rely on linear or rule-based approaches.

Leveraging self-regulatory learning, Zollanvari et al. (2017) developed a novel machine learning to construct and validate a CPGA predictive model [31]. The model could detect interventions for assisting students with a low GPA. Also, Alsalman et al. (2019) developed a decision tree and ANN to calculate students' academic performance by predicting the expected GPA [32]. Following this, a predictive learning model was established for the early adoption of the at-risk student in a course [33]. Finally, Lykourentzou et al. (2009) devised an early and dynamic student final achievement prediction method to cluster them in two virtual groups based on their performance in e-learning courses [34]. They found the accurate prediction possible while still in the early stages. This model improved educational services and offered customized assistance based on their predicted performance levels. Using the logistic regression model, Raju and Schumacker (2015) used the earliest possible student data to explore students' retention characteristics leading to graduation at HEIs [33]. Therefore, it is evident that future HEIs need to adopt these models to serve students better.

Romero et al. (2013) demonstrated that students' final marks could be predicted based on their involvement in on-line discussion forums. Though they indicate that this early detection before the end of the course is critical, it is less accurate [35]. Then, Hu, Lo and Shih (2014) developed early warning systems for predicting the "at-risk" students while progressing through an online course [36]. Using data mining techniques to evaluate the learning portfolios from a fully on-line course showed that CART supplemented by AdaBoost performed better. Finally, Gray, McGuinness and Owende (2014) reported a classification model that identifies college students at risk of failing during their first year of study based on age, gender, prior academic performance, and psychometric indicators collected from three student cohorts [37]. This study constitutes a critical predictive learning model application for predicting learner progression at the tertiary level.

Notably, instead of evaluating instructors' course performance through a questionnaire based on student perceptions, Agaoglu (2016) used data mining techniques to develop a model for predicting instructor performance [28]. The results show that an instructor's success mainly depends on the students' perceptions and interests. Moreover, Xu, Moon and Van Der Schaar (2017) reported a machine-learning model to track and predict student performance in a degree program. An essential pedagogical mediation is to predict students' future performance in a graduate course based on their continuous assessment records. This model allows HEIs to determine students who will complete their studies in record time [29]. However, all these can be achieved if the staff is upskilled to use the technology and available data to personalize students' experiences. This demonstrates that students are the main denominator, and universities must be mindful of this and continuously innovate. Thus, future universities must constantly evolve to remain relevant by considering what is happening in society.

The systems thinking approach has been studied and applied in HEIs. Systems thinking is "a holistic approach that enables simultaneous analysis of the parts and the whole itself, their evolution, overlaps, and dynamic interactions" [38]. Systems thinking as used in learning seeks to discover how eruptive technologies are integrated to improve student achievement and organizational success. Therefore, when thinking of learning systems, it is critical to establish the system's key entity and function and organize the education to attain the best possible learning outcomes [39]. The student constitutes the key entity, while the key function involves planning and providing materials for the student to achieve the required competence. Some adjustments are made in the learning materials when competence is not attained. Most significantly, ongoing conversations are in the process between the learner and instructor to ensure that the learner masters the learning task. This demonstrates that systems thinking implicitly achieves student-centered learning, which shifts the focus of instruction from the teacher to the student. This forms one of the main goals of the university of the future.

Dhukaram et al. (2018) investigate the complexities involved in HEIs and how to use the systems thinking approach to understand the education ecosystem. At HEIs, systems thinking is applied to develop courses or curricula, think about and effect change in the organization, and reflect on or research the nature of learning [40]. Moving from passive to active learning strategies required by future universities has increased the complexity. Therefore, it offers means to assist in responding to expanding the delivery complexities from traditional models to more adaptive models required at future universities. Besides, HEIs are inherently complex because they constitute various components such as hardware, software, instructors, students, delivery, management and procedures, organization and regulations, and government [41]. Therefore, the systems thinking approach is necessary to find how interactions between the people, technology, and the organization achieve personalized learning, improve student and instructor experiences, and bridge the skills gap between the university and the workplace.

3 Methodology

We conduct a landscape analysis of PLA and the systems thinking approach based on a systematic review of recent literature. We collect and summarize information on PLA use and identify how it is applied and expected to benefit learners, educators during the course, and HEIs. While several reviews appear in the literature, we adopt a system thinking approach and investigate factors and interactions to improve student and organizational success. Understanding systems allows examining how one action to a part of a system can affect the entire system, thus deepening and enhancing learning. Therefore, applying several search terms and keywords related to PLA techniques at HEIs, we conducted searches and collected the most relevant manuscripts containing information pertinent to our goal.

4 Impact of Systems Thinking Approach and Predictive Analytics on HEI Stakeholders

Systems thinking uses various tools and cognitive frameworks to improve understanding of complex behaviors or phenomena within systems, occurring naturally or artificially and from a holistic perspective [42]. It is a powerful approach for understanding why situations are the way they are and how to improve results. Systems thinking demonstrates the importance of intelligently integrated technologies to enable HEIs stakeholders to design, track, and evaluate programmatic interventions to improve student achievement and organizational success [43]. For example, systems thinking and PLA allow us to establish which students require attention at a specific time. Based on data, it provides feedback about whether the course has been appropriately designed. Systems thinking allows education leaders to recognize relationships between different components and use them to solve problems. It provides a practical approach to diagnosing a problem so that solutions addressing the needs of a particular stakeholder can be identified [44]. The systematic review provides an integrative report concerning methods, benefits, and challenges of PLA and how to apply it more effectively to improve teaching and learning at HEIs.

Though the PLA research field is still emerging, it is rapidly growing. It has made exceptional progress as researchers and HEIs increasingly understand the PLA's potential in supporting the learning process. Some benefits include predicting students' performance and retention, detecting undesirable learning behaviors and emotional states, and learning patterns, identifying at-risk students, and making an immediate follow-up. However, some challenges involve data tracking, collection, evaluation, analysis, optimizing learning conditions, and ethics and privacy issues. According to Avella et al. (2016), the challenge with the rapid embrace of these data analytic techniques is that they divert educators' attention [45]. Moreover, Tsai and Gasevic (2017) highlight the need to establish communication channels between or among stakeholders and adopt teaching practices based on learning analytics [46].

To avoid these challenges, instead of simply discussing the effects of PLA in HEIs, we leverage the systems thinking approach to study its implications in HEIs. Considering that systems thinking is holistic rather than reductionist, we consider systems more than the sum of their parts. Therefore, systems thinking requires the complete educational system to be coordinated and must focus simultaneously on the educational system's main areas, which is a significant challenge for instructors.

4.1 Learner

Learners constitute one of the principal stakeholders of HEIs. Systems thinking, PLA, and data mining approaches could bring numerous positive effects and improve learner performance if effectively implemented. Since PLA only focuses on the individual learner instead of the entire learning program; therefore, the latter can be accommodated through the systems thinking approach. This makes a combination of PLA and systems thinking uniquely valuable for addressing ineffective learning challenges. Ineffective learning happens if the course content is not explicit, there is an opportunity to apply new knowledge, a lack of institutional support, and a possibility of misalignment of goals and priorities. Thus, PLA allows the instructor to determine a student's understanding of the teaching material and how they will apply the knowledge to solve assignment, tutorial, and examination practice questions. Notably, this can be addressed by adopting adaptive learning. Adaptive learning courseware allows modification of students' learning routes based on the student's interactions with technology. Systems thinking will assist the instructor in precisely identifying a student's shortcomings and customizing their academic experience. These tools allow students to accelerate their learning journey by moving quickly through the course curriculum they already know and providing additional support. Moreover, systems thinking assists address complex multidimensional problems and interdisciplinary teaching and learning. Therefore, encouraging students to think from a systems perspective allows problem-solving outside their discipline-based channels.

PLA and systems thinking approach empower students by informing them when they are at risk due to not submitting a course assessment. A PLA model will predict that the student needs to act or continue being on course. One may design a dashboard for a particular course such that when a student login the system, they will see some warning signs, for instance, a red light meaning "at-risk" and a green light indicating "on-track." This will keep students informed about their progress and enable them to make necessary adjustments accordingly. This dashboard can be customized such that it appears on the educator's side to track the progress and performance of students. Also, if learners display signs of ineffective learning, PLA tools can send notifications to educators to monitor learner progress and watch for indicators when the learning is not being applied. While these indicators depend on the type of higher education system and mode of teaching, key indicators include GPA, clickstream data, learning history, task completion rates, student's progress, when a user was last logged in, number of answer attempts, and learning path data. An educator can use the low quiz score and participation as an indicator to predict the learner's performance. Also, the educator can compare a specific metric against what they usually mean for course performance. For example, consistent low quiz scores and absence from forum participation may show a student who is not actively engaged. This information enables educators to identify prospects for intervention. Therefore, HEI can use PLA to identify actionable metrics for learners and could assist them in mapping the course for their learning. Since students have access to their data, the process becomes transparent and encouraging. Therefore, this openness and transparency are critical in breaking down siloes and building a sustainable future university.

While PLA offers numerous opportunities to discover insightful information about students at HEIs, certain attitudes and approaches must guide the usage, processing, and presentation of analytics results based on person-related data. Data relating to students may be divided into two categories: static and dynamic student data. Numerous students are willing to provide their data in exchange for more effective, personalized, and support services [13]. Though students know the benefits and risks of being ingrained in data exchange, a challenge lies in deciding whether institutions must continue to exchange their data for more effective and supported learning experiences [15]. Moreover, challenges arise if there are chances that the collected data may restrain their future studies or employment opportunities [47]. Fortunately, legal and regulatory developments have been developed to protect personal privacy, particularly students' right to data privacy [48]. However, in most cases, students do not control the type of data collected and how it will be used; they delegate their trust and transparency to their HEIs [49]. Therefore, HEIs must balance legitimate institutional interests and student benefits by exercising appropriate transparency.

4.2 Courses and Learning Design

Systems thinking may be explicitly applied at HEIs to develop courses or learning support processes that enable students to use a systems approach for managing complex scenarios and develop training or education programs or curricula for students in the sector. On the other hand, PLA models could be powerful tools that provide fundamental insights for improving course design. These models offer a clearer picture of how course learning outcomes influence learners' performance and subsequent results. This can be determined based on learner ratings on specific modules, activity drop-off rates, and weekly participation metrics. For instance, a continued activity drop-off might indicate a lack of engagement with the material or difficulty in the course material. So, educators could monitor learning performance and refine upcoming iterations concerning course content. This demonstrates that educators can increase the effectiveness of course designs through PLA. Other indicators or metrics that might be used are student's performance relative to the total enrolled, time spent per course, which course parts learners get low or high scores, learning course data, and course rating. However, these indicators may not immediately benefit from providing quick and accurate insights on how to improve course design.

Using learning resources, interactions, and learners' participation, PLA provides means to consolidate individual indicators into an analytics report. For instance, the dashboard must deliver analytical reports in real time: enrollment status, learner activity, learner progress, and potential problems. The enrollment status must provide information about the number of enrolled students and give information on who is passing or failing the course and could immediately be assisted. In addition, students can identify courses to register for in the next academic semester through program recommender systems. The learner activity gives a detailed account of learner interactions and patterns, including information such as materials they interacted with the most or omitted course. The learner's progress informs the educator where learners are in their learning journey, while the potential problems report analyses learner assessment data. This report will also notify the educator if there are course materials that learners fail to grasp.

Furthermore, it provides an instructional approach where the course design (i.e., curriculum, content, format, and delivery method) is optimized according to each learner's needs, abilities, and preferred learning mode (i.e., offers personalized learning). Most significantly, this is an essential characteristic of a future university. For instance, a larger group of students may come from different parts of the country and possess different skill levels (i.e., science, engineering, art, and humanities), which is critical in the future of work. Thus, PLA and AI ensure an educator has an option to implement personalized learning and provides them with multiple directions for future delivery of the course material. Furthermore, it offers an opportunity for self-regulation when the educator provides feedback to learners to make corrections where they are going wrong. The personalized learning platform could integrate the online interactive interface to collate learners' feedback. Learners can share their course experience and queries or request refreshers where necessary.

Besides PLA being a notable advance at HEIs and still promises numerous potential benefits for stakeholders by unlocking value from data, the field faces challenges [50]. Several institutions are not ready to adopt PLA data [51, 52]. For instance, in some institutions, educators who are supposed to be at the forefront of implementing PLA models are not prepared to adopt them. Furthermore, educators lacking a technical background find PLA data and visualizations challenging to understand. This makes it a challenge if educators are tasked to interpret the result or visualizations. An educator's failure to interpret PLA data's output implies that appropriate action or intervention strategies may not be taken or take incorrect intervention strategies. However, PLA and systems thinking approach offers numerous opportunities to empower educators to monitor and intervene with their students effectively compared to other methods and improve learning outcomes. These outcomes must significantly aim to bridge the increasing skills gap between education and an increasingly competitive entrepreneurial and work environment.

4.3 Institutions

PLA and systems thinking approach improve decision-making, inform resource allocation, and increase productivity at the institutional level by harnessing unprecedented large amounts of data collected by digital technologies [46]. HEIs can use PLA to identify actionable indicators for learners to assist them in mapping the course for their learning. Taking appropriate action or remedy at each learning stage will improve the overall learner's performance, progression, experience and satisfaction, teaching quality and innovations, and the overall institutional performance, including ranking. Therefore, institutions need to adopt PLA and systems thinking approach to improve student retention and academic success, inform curriculum design and learning support, and personalized learning systems and course recommendations [45]. While institutions have addressed and managed some PLA goals, progression and retaining students have been challenging [53]. Moreover, institutions can leverage PLA and systems thinking to shift institutional culture, though this is not a linear process and requires complicated tools. Tsai et al. (2020) discuss some of the complex issues that impede the scaling of learning analytics and provide approaches by HEIs and stakeholders to address them [54].

According to Tsai et al. (2020) and Waheed et al. (2018), the inception of PLA by various countries has been alarming, though still embryonic among the least developed countries [46]. The USA, Spain, UK, Australia, Germany, and Canada are leading the adoption and implementation of PLA and the systems thinking approach [55]. The goal is to measure and improve student and organizational efficiency and inform decision-making [56]. Several institutions have pioneered academic analytics' initiatives by using predictive tools. For example, Purdue University leverages predictive modeling using data from their LMS (Course Signals) to provide immediate feedback to students regarding their performance. Also, the system provides early intervention and frequent and ongoing feedback [57, 58].

The University of Alabama has developed a predictive model using data files from first-year students to study student retention based on several indicators like English course grades, the sum of hours earned, and learners' demographics [59]. Moreover, Northern Arizona University developed a Grade Performance System (GPS), an early warning alert, and retention system that provides students with grades, attendance, academic issues, and positive feedback. The system also provides students options and resources depending on the alert's nature [60, 61]. In addition, the Rio Salado Community College in Arizona uses the PACE (Progress and Course Engagement) system for an automated, systematic early-alert system allowing instructors to launch proactive interventions at any time during the course to help students who display signs of struggling as well as providing interventions where necessary [60, 62].

The Graduate School of Medicine at the University of Wollongong uses an LCMS, Equella, to collect information and data about clinical placements throughout medical school training. Students use the analytics tool to record their curriculum experiences. The tool tracks the student's level of involvement during their rotation assignment to the curriculum [63]. At the University of Maryland Baltimore Country, they use Blackboard LCMS to collect and manage data and information about students as part of an integrated strategy for teaching, learning, and technology [64]. The system can identify student learning outcomes, infrastructure support, and online or hybrid strategy. In addition, the system can examine individual student activity and grade distribution. If the grade falls below a certain level, it displays a warning to the student and educator. The University of Michigan uses E2Coach and Michigan Tailoring System (MTS) to provide tailored and personalized recommendations [64]. At Sinclair Community College, they developed the Student Success Plan (SSP), which integrates several variables [61, 65]. At Baylor University, they created an enrollment predictive modeling tool that gathers and analyses large amounts of prospective students' data to enhance a complex admissions strategy. The predictive model scores become part of the database, which admission counselors query to identify prospective students. The model is such that no mailings go to prospective students with low-scoring inquiries as they are least likely to enroll [66]. This admission strategy impacts future cohorts.

Based on the above discussion, several measures are necessary to maximize the benefits of PLA and the systems thinking approach. Therefore, institutions must have a robust Student Management System (SMS) configured to collect essential student data. The educator must understand how students learn best and struggle based on this data and appropriate tools. Educators must take advantage of the collected data to understand learners' behaviors, preferences, successes, and failures. Proper data can effectively improve course design, especially monitoring the learning process, dis-covering student patterns, and exploring student data. It can also be used to find early indicators for student success or failure, assess course materials' effectiveness, intervenes, supervises, advises, assists students, and improve course delivery and the course environment. Moreover, the educator will know which course students face challenges, which course section prompts questioning, and learners have challenges with assessments. Therefore, the more the educators know and understand students, the better they can redesign the course and meet their needs.

5 Discussion

5.1 Challenges of Predictive Learning Analytics and Systems Thinking Approach

Despite opportunities and success presented by PLA and the systems thinking approach, these have not been widely adopted due to resource demands and social complexities [46]. Other challenges embrace data tracking, collection, evaluation, and analysis, optimizing learning conditions, ethics, and privacy concerns [48]. Concerning these concerns, stakeholders need to find effective ways to effectively deploy data analytics tools and techniques to improve teaching and learning in higher education. Though this is impressive, the technology to deliver this potential is still emerging, and more work is still outstanding, especially in understanding the pedagogical usefulness of PLA [66]. Therefore, institutions and stakeholders have a critical role in gathering meaningful data, achieving transparency, and designing unbiased algorithms that inform processes and good practices. These requirements are not exhaustive; slightly misusing intelligent tools could endanger various stakeholders. For instance, there is a risk that profiling issues may arise during classifications of learners who meet specific characteristics, such as GPA [67]. While students, educators, and institutions have inherent expectations, the concern is that PLA may add a set of data-driven expectations. Data privacy, ethics, transparency, and use pose several problems, especially for students [15]. According to Pardo and Siemens (2014), data collection for PLA has brought concerns about intruding the learner privacy [18]. Thus, there is a tradeoff between optimizing what data can be collected or done and ensuring its responsible use [68, 69]. Furthermore, it is still unclear which rights learners possess on their data and the degree to which they are accountable for acting on the recommendations supplied by PLA.

Watters (2012) raised concerns about whether HEIs measure student learning or boost their retention and completion of a course. When we consider different data types for learning analytics, for instance, the total course tools accessed in a student LMS, or the sum of posts "read" or "likes" in discussion forums. It is not easy to establish whether these are still the proxy measures of learning. However, this does not mean PLA can-not boost learning, but there needs to be clarity about what is measured and predicted. Thus, we can apply system thinking to examine the interaction, and feedback loops produce outcomes. Therefore, this allows diagnosis of the problem to identify solutions that address the needs.

The integration, deployment, and intelligent tools require advanced skillsets and financial investment [56]. This requires substantial strategic investment in both human capital development (hiring new talent, upskilling existing staff), changing the work model (introducing new policies, management, and processes), and related infrastructure (purchasing new computing equipment and insurance). These challenges may become pronounced if competing priorities are already in progress [46]. This provides an opportunity for systems thinking to establish dynamic relationships between various activities and expected synergies between different sectors or departments of the organization. Thus, this will present an opportunity for enhanced collaboration and innovation. Considering these challenges and possibilities in implementing PLA, there is a need for stakeholder engagement and a shared vision. However, unequal engagement with critical stakeholders may lead to inertia and the project's collapse. Therefore, a co-design process for developing tools, strategies, and policies is essential where foreseeable.

5.2 Opportunities with PLA and Systems Thinking

The LMS and application can monitor the times an individual logs into the course room. Using these technologies in learning is seen chiefly as resource integration or a new procedure for facilitating learning activities and outcomes. However, these platforms also provide substantial documentation to establish a learner's involvement in the course upon login. This ability to track data gives valuable information to planning and implementing new educational programs. Therefore, tracking reveals students' level of engagement with the curriculum and detects the course material, which may confuse them. This information provides critical input for planning and implementing the learning course and design to continue implementing the educational program.

5.3 Limitations of the Study

The limitations of this work could be linked closely to limitations of the research method, i.e., the limitations of a systematic review. First, in some cases, the summary presented in a systematic review and the literature's meta-analysis is only reliable as the procedure for estimating the effect in each primary study. This means conducting a meta-analysis fails to overcome inherent challenges due to the design and execution of preliminary investigations. Moreover, it fails to correct biases resulting from the selective publication or publication bias [70]. Furthermore, studies with statistically significant results will probably get published compared to non-significant results. Therefore, meta-analyses exclusively based on published literature will likely produce biased results [71]. Moreover, with the systematic review, the inspection of article references is mainly used in identifying additional relevant articles. The challenge of this procedure is that citing a previous work may be far from objective and learning the literature through scanning the reference lists may lead to a biased sample of studies. On the other hand, there is the likelihood of over-citing unsupportive studies. Furthermore, merely labeling a manuscript as a "systematic review" does not promise that a review was carried out or reported with diligence. For instance, the biases in selecting and assessing the literature and assumptions are usually unknown.

6 Conclusion

This work provided a systematic literature review about complementing systems thinking and PLA techniques to create sustainable future universities. It suggests that adopting a perspective system to problem situations provides a more helpful framework for representing the real world than the reductionist, discipline-based one. We demonstrated how these techniques could unlock actionable intelligence from statistical analysis of student data combined with other traditional data sources like demographics or academic success to make actionable decisions regarding students, learning and course design, and institutions. Therefore, after identifying students with academic difficulties, educators and course advisors can customize learning paths or provide instruction tailored to specific students' learning needs. This is one of the goals or characteristics of future universities. Furthermore, we applied the systems thinking approach to establish interconnectedness between the various aspects and assist in framing complex problems, often misdiagnosed when using linear thinking, which is critical for future universities toward solving real-world problems.

In addition to contributing to the conceptual and theoretical understanding of PLA and its relevance within HEIs, we identified opportunities and challenges associated with exploring and implementing analytical techniques. These challenges and opportunities are linked directly to establishing and sustaining future universities. We also described how PLA could inform learning and course design, learning programs, and teaching and institutional policies. While HEIs continue to realize systems that collect massive and varied data, several information and technology units play a critical role in supporting analytics. This implies that staff, including educators, will have more than traditional information and technology skills. Accordingly, the educator must be proficient at mining data, understanding its nature, creating metadata to provide long-term data management, analyzing and visualizing data from different sources, and developing models for actionable decision-making.

As a result, HEIs are recommended to focus on tools and systems and expertise, process, and policies to achieve a sustainable university of the future. Institutions must build their human resources capacity and scale up the implementation of digital technology activities. Also, they must plan for infrastructure that supports data analytics across departments in the institution. However, this may require additional financial resources and a changed-management approach to provide a framework for creating a conducive climate of transformative change. Through collaborating and sharing common data across different departments, data analytics can help institutions bridge academic affairs and student affairs toward advancing and informing learning. This opens more collaborative opportunities for research the institution and potential collaborations with the industry.

References

1. Herodotou, C., Rienties, B., Boroowa, A., Zdrahal, Z., & Hlosta, M. (2019). A large-scale implementation of predictive learning analytics in higher education: the teachers' role and perspective. *Educational Technology Research and Development, 67*(5), 1273–1306.
2. Yakubu, M. N., & Abubakar, A. M. (2021). *Applying machine learning approach to predict student' performance in higher educational institutions.* Kybernetes.

3. Conole, G., De Laat, M., Dillon, T., & Darby, J. (2008). 'Disruptive technologies', 'pedagogical innovation': What's new? Findings from an in-depth study of students' use and perception of technology. *Computers & Education, 50*(2), 511–524.
4. Dyckhoff, A. L., Zielke, D., Bültmann, M., Chatti, M. A., & Schroeder, U. (2012). Design and implementation of a learning analytics toolkit for teachers. *Educational Technology & Society, 15*(3), 58–76.
5. Durak, G., & Çankaya, S. (2019). Learning management systems: Popular LMSs and their comparison. In *Handbook of research on challenges and opportunities in launching a technology-driven international university* (pp. 299–320). IGI Global.
6. Kraleva, R., Sabani, M., & Kralev, V. (2019). An analysis of some learning management systems. *International Journal on Advanced Science, Engineering and Information Technology, 9*(4), 1190–1198.
7. York, S., Lavi, R., Dori, Y. J., & Orgill, M. (2019). Applications of systems thinking in STEM education. *Journal of Chemical Education, 96*(12), 2742–2751.
8. Herodotou, C., Naydenova, G., Boroowa, A., Gilmour, A., & Rienties, B. (2020). How can predictive learning analytics and motivational interventions increase student retention and enhance administrative support in distance education? *Journal of Learning Analytics, 7*(2), 72–83.
9. Norris, D., Baer, L., Leonard, J., Pugliese, L., & Lefrere, P. (2008). Action analytics: Measuring and improving performance that matters in higher education. *Educause Review, 43*(1), 42.
10. Slade, S., Prinsloo, P., & Khalil, M. (2019). Learning analytics at the intersections of student trust, disclosure and benefit. In *Proceedings of the 9th international conference on learning analytics & knowledge* (pp. 235–244).
11. Schreiner, L. A. (2009). *Linking student satisfaction and retention*. Noel-Levitz.
12. Sarker, M. N. I., Wu, M., Cao, Q., Alam, G. M., & Li, D. (2019). Leveraging digital technology for better learning and education: a systematic literature review. *International Journal of Information and Education Technology, 9*(7), 453–461.
13. Larrabee Sønderlund, A., Hughes, E., & Smith, J. (2019). The efficacy of learning analytics interventions in higher education: A systematic review. *British Journal of Educational Technology, 50*(5), 2594–2618.
14. Drachsler, H., Hoel, T., Scheffel, M., Kismihók, G., Berg, A., Ferguson, R., & Manderveld, J. (2015). Ethical and privacy issues in the application of learning analytics. In *Proceedings of the fifth international conference on learning analytics and knowledge* (pp. 390–391).
15. Slade, S., & Prinsloo, P. (2013). Learning analytics: ethical issues and dilemmas. *American Behavioral Scientist, 57*(10), 1509–1528.
16. Rubel, A., & Jones, K. M. (2016). Student privacy in learning analytics: An information ethics perspective. *The Information Society, 32*(2), 143–159.
17. Agasisti, T., & Bowers, A. J. (2017). Data analytics and decision making in education: Towards the educational data scientist as a key actor in schools and higher education institutions. In *Handbook of contemporary education economics*. Edward Elgar Publishing.
18. Pardo, A., & Siemens, G. (2014). Ethical and privacy principles for learning analytics. *British Journal of Educational Technology, 45*(3), 438–450.
19. Walczak, S. (1994, June). Categorizing university student applicants with neural networks. In *Proceedings of 1994 IEEE international conference on neural networks (ICNN'94)* (Vol. 6, pp. 3680–3685).
20. Gottlieb, E. (2001). Using integer programming to guide college admissions decisions: A preliminary report. *Journal of Computing Sciences in Colleges, 17*(2), 271–279.
21. Fong, S., Si, Y. W., & Biuk-Aghai, R. P. (2009). Applying a hybrid model of neural network and decision tree classifier for predicting university admission. In *2009 7th international conference on information, communications and signal processing (ICICS)* (pp. 1–5).
22. Ragab, A. H. M., Mashat, A. F. S., & Khedra, A. M. (2012, November). HRSPCA: Hybrid recommender system for predicting college admission. In *2012 12th international conference on intelligent systems design and applications (ISDA)* (pp. 107–113).
23. Huang, S., & Fang, N. (2013). Predicting student academic performance in an engineering dynamics course: A comparison of four types of predictive mathematical models. *Computers & Education, 61*, 133–145.

24. Xia, F., Wang, W., Bekele, T. M., & Liu, H. (2017). Big scholarly data: A survey. *IEEE Transactions on Big Data, 3*(1), 18–35.
25. Scheffel, M., Drachsler, H., Toisoul, C., Ternier, S., & Specht, M. (2017, September). The proof of the pudding: examining validity and reliability of the evaluation framework for learning analytics. In *European conference on technology enhanced learning* (pp. 194–208). Springer.
26. Conijn, R., Van den Beemt, A., & Cuijpers, P. (2018). Predicting student performance in a blended MOOC. *Journal of Computer Assisted Learning, 34*(5), 615–628.
27. Hassan, H., Anuar, S., & Ahmad, N. B. (2019, May). Students' performance prediction model using meta-classifier approach. In *International conference on engineering applications of neural networks* (pp. 221–231). Springer.
28. Athani, S. S., Kodli, S. A., Banavasi, M. N., & Hiremath, P. S. (2017). Student academic performance and social behavior predictor using data mining techniques. In *2017 international conference on computing, communication and automation (ICCCA)* (pp. 170–174).
29. Tan, M., & Shao, P. (2015). Prediction of student dropout in e-Learning program through the use of machine learning method. *International Journal of Emerging Technologies in Learning, 10*(1).
30. Elçi, A., & Abubakar, A. M. (2021). The configurational effects of task-technology fit, technology-induced engagement and motivation on learning performance during Covid-19 pandemic: An fsQCA approach. *Education and Information Technologies*, 1–19.
31. Zollanvari, A., Kizilirmak, R. C., Kho, Y. H., & Hernández-Torrano, D. (2017). Predicting students' GPA and developing intervention strategies based on self-regulatory learning behaviors. *IEEE Access, 5*, 23792–23802.
32. Alsalman, Y. S., Halemah, N. K. A., AlNagi, E. S., & Salameh, W. (2019). Using decision tree and artificial neural network to predict students academic performance. In *2019 10th international conference on information and communication systems (ICICS)* (pp. 104–109).
33. Marbouti, F., Diefes-Dux, H. A., & Madhavan, K. (2016). Models for early prediction of at-risk students in a course using standards-based grading. *Computers & Education, 103*, 1–15.
34. Lykourentzou, I., Giannoukos, I., Mpardis, G., Nikolopoulos, V., & Loumos, V. (2009). Early and dynamic student achievement prediction in e-learning courses using neural networks. *Journal of the American Society for Information Science and Technology, 60*(2), 372–380.
35. Romero, C., López, M. I., Luna, J. M., & Ventura, S. (2013). Predicting students' final performance from participation in on-line discussion forums. *Computers & Education, 68*, 458–472.
36. Hu, Y. H., Lo, C. L., & Shih, S. P. (2014). Developing early warning systems to predict students' online learning performance. *Computers in Human Behavior, 36*, 469–478.
37. Gray, G., McGuinness, C., & Owende, P. (2014). An application of classification models to predict learner progression in tertiary education. In *2014 IEEE international advance computing conference (IACC)* (pp. 549–554). IEEE.
38. Michalopoulou, E., Shallcross, D. E., Atkins, E., Tierney, A., Norman, N. C., Preist, C., et al. (2019). The end of simple problems: repositioning chemistry in higher education and society using a systems thinking approach and the united nations' sustainable development goals as a framework. *Journal of Chemical Education, 96*(12), 2825–2835.
39. Banathy, B. H. (1999). Systems thinking in higher education: Learning comes to focus. *Systems Research and Behavioral Science: The Official Journal of the International Federation for Systems Research, 16*(2), 133–145.
40. Dhukaram, A. V., Sgouropoulou, C., Feldman, G., & Amini, A. (2018). Higher education provision using systems thinking approach–case studies. *European Journal of Engineering Education, 43*(1), 3–25.
41. Ison, R. (1999). Applying systems thinking to higher education. *Systems Research and Behavioral Science: The Official Journal of the International Federation for Systems Research, 16*(2), 107–112.
42. Senge, P. M. (2006). *The fifth discipline: The art and practice of the learning organization.* Currency.
43. Orgill, M., York, S., & MacKellar, J. (2019). Introduction to systems thinking for the chemistry education community. *Journal of Chemical Education, 96*(12), 2720–2729.

44. Raju, D., & Schumacker, R. (2015). Exploring student characteristics of retention that lead to graduation in higher education using data mining models. *Journal of College Student Retention: Research, Theory & Practice, 16*(4), 563–591.
45. Avella, J. T., Kebritchi, M., Nunn, S. G., & Kanai, T. (2016). Learning analytics methods, benefits, and challenges in higher education: A systematic literature review. *Online Learning, 20*(2), 13–29.
46. Tsai, Y. S., Rates, D., Moreno-Marcos, P. M., Muñoz-Merino, P. J., Jivet, I., Scheffel, M., et al. (2020). Learning analytics in European higher education—Trends and barriers. *Computers & Education, 155*, 103933.
47. Hauff, S., Veit, D., & Tuunainen, V. (2015). *Towards a taxonomy of perceived consequences of privacy-invasive practices*.
48. Ferguson, R. (2019). Ethical challenges for learning analytics. *Journal of Learning Analytics, 6*(3), 25–30.
49. Hauff, S., Trenz, M., Tuunainen, V. K., & Veit, D. (2016). *Perceived threats of privacy invasions: Measuring privacy risks*.
50. Martin, F., & Ndoye, A. (2016). Using learning analytics to assess student learning in online courses. *Journal of University Teaching & Learning Practice, 13*(3), 7.
51. Lockyer, L., Heathcote, E., & Dawson, S. (2013). Informing pedagogical action: Aligning learning analytics with learning design. *American Behavioral Scientist, 57*(10), 1439–1459.
52. Tempelaar, D. T., Rienties, B., & Giesbers, B. (2015). In search for the most informative data for feedback generation: Learning analytics in a data-rich context. *Computers in Human Behavior, 47*, 157–167.
53. Thomas, L. (2016). Developing inclusive learning to improve the engagement, belonging, retention, and success of students from diverse groups. In *Widening higher education participation* (pp. 135–159). Chandos Publishing.
54. Tsai, Y. S., Whitelock-Wainwright, A., & Gašević, D. (2020). The privacy paradox and its implications for learning analytics. In *Proceedings of the tenth international conference on learning analytics & knowledge* (pp. 230–239).
55. Waheed, H., Hassan, S.-U., Aljohani, N. R., & Wasif, M. (2018). A bibliometric perspective of learning analytics research landscape. *Behaviour & Information Technology, 37*(10–11), 941–957.
56. Arroway, P., Morgan, G., O'Keefe, M., & Yanosky, R. (2016). *Learning analytics in higher education* (Technical report). EDUCAUSE Center for Analysis and Research. https://library.educause.edu/~/media/files/library/2016/2/ers1504la.pdf
57. Essa, A., & Ayad, H. (2012, April). Student success system: risk analytics and data visualization using ensembles of predictive models. In *Proceedings of the 2nd international conference on learning analytics and knowledge* (pp. 158–161).
58. Arnold, K. E., & Pistilli, M. D. (2012, April). Course signals at Purdue: Using learning analytics to increase student success. In *Proceedings of the 2nd international conference on learning analytics and knowledge* (pp. 267–270).
59. Dietz-Uhler, B., & Hurn, J. E. (2013). Using learning analytics to predict (and improve) student success: A faculty perspective. *Journal of Interactive Online Learning, 12*(1), 17–26.
60. Picciano, A. G. (2014). Big data and learning analytics in blended learning environments: Benefits and concerns. *IJIMAI, 2*(7), 35–43.
61. Smith, V. C., Lange, A., & Huston, D. R. (2012). Predictive modeling to forecast student outcomes and drive effective interventions in online community college courses. *Journal of Asynchronous Learning Networks, 16*(3), 51–61.
62. Olmos, M., & Corrin, L. (2012). Academic analytics in a medical curriculum: Enabling educational excellence. *Australasian Journal of Educational Technology, 28*(1).
63. Mattingly, K. D., Rice, M. C., & Berge, Z. L. (2012). Learning analytics as a tool for closing the assessment loop in higher education. *Knowledge Management & E-Learning: An International Journal, 4*(3), 236–247.
64. Fletcher, J., Grant, M. N., Karp, M. J. M., & Ramos, M. (2016). *Integrated planning and advising for student success (iPASS): State of the literature*.

65. Van Barneveld, A., Arnold, K. E., & Campbell, J. P. (2012). Analytics in higher education: Establishing a common language. *EDUCAUSE Learning Initiative, 1*(1), 1–11.
66. Johnson, L., Adams, S., & Cummins, M. (2012). *The NMC horizon report: 2012 Higher education edition*. The New Media Consortium.
67. Van Barneveld, A., Arnold, K. E., & Campbell, J. P. (2012). Analytics in higher education: Establishing a common language. *EDUCAUSE Learning Initiative, 1*(1), 1-ll.
68. Tsai, Y. S., Perrotta, C., & Gašević, D. (2020). Empowering learners with personalised learning approaches? Agency, equity and transparency in the context of learning analytics. *Assessment & Evaluation in Higher Education, 45*(4), 554–567.
69. Palma, S., & Delgado-Rodriguez, M. (2005). Assessment of publication bias in meta-analyses of cardiovascular diseases. *Journal of Epidemiology & Community Health, 59*(10), 864–869.
70. Jadad, A. R., Moher, M., Browman, G. P., Booker, L., Sigouin, C., Fuentes, M., & Stevens, R. (2000). Systematic reviews and meta-analyses on treatment of asthma: Critical evaluation. *BMJ, 320*(7234), 537–540.
71. Viberg, O., Hatakka, M., Bälter, O., & Mavroudi, A. (2018). The current landscape of learning analytics in higher education. *Computers in Human Behavior, 89*, 98–110.

Reimaging Continuing Professional Development in Higher Education – Toward Sustainability

Saba Mansoor Qadhi and Haya Al-Thani

1 Introduction

The United Nations General Assembly approved 17 Sustainable Development Goals (SDGs) in 2015 in the hope of attaining them by 2030. The seventeen objectives seek to "ensure a sustainable, secure, successful, and inclusive living on Earth for now as well as for the future" [1]. According to UNESCO, education is a significant tool for accomplishing the sustainable development goals (SDGs) since it increases learning, abilities, beliefs, perspectives, critical reasoning, competencies, systemic thinking, and accountability and empowers subsequent generations to achieve required transformative differences in the world as well [2, 3]. Education for sustainable development is defined by UNESCO as a foundation of strength for students to "take intelligent choices and accountable actions for ecological responsibility, financial feasibility, and a just community, for generations to come, while preserving ethnic diversity" [4]. This is because sustainability may be regarded as a task for humanity and a struggle to learn how to deal more responsibly, as per [5]. In this respect, continuing professional development (CPD) is largely viewed as an extremely important concept in the aspect of higher education, contributing to students' personal and professional growth as well as the progress in education and learning. Professional development is intimately tied to individuals' daily tasks and must be integrated into a larger system of ongoing learning [6]. It refers to the expansion and maturity of a persons' knowledge, abilities, and dispositions because of public and private activities of

S. M. Qadhi (✉)
Qatar University, Doha, Qatar
e-mail: sabaa@qu.edu.qa

H. Al-Thani
Museum of Qatar, Doha, Qatar

learning [7]. However, sustainability in education is not confined to climate change and promoting green technology only.

To ensure that leaders in education are supported and developed internally, sustainability in higher education must emphasize the role of leadership and prevent "leadership waste" while enhancing organizational effectiveness [8]. Therefore, sustainability-based education has an impact on educational curriculum, as well as the accompanying processes, and outcomes [9]. It will enable the emergence of an even more sustainable leadership model that focuses on relational leadership models and highly personalized planning processes, resulting in enhanced leadership practices and, as a result, increased research, and education activities at HE institutions. HE institutions can serve as a bridge between a variety of stakeholders since universities and colleges bear a special duty for educating and empowering professionals, as well as the disseminating of opinions, and resources [10]. Therefore, this book chapter aims to identify, explore, and describe the role of leadership in the context of sustainable education within the HE context to ensure the support and development of leaders that ensure increased research, learning, and teaching in HE institutions. The objectives of this book chapter are as follows:

- To discuss the role and importance of CPD in higher education.
- To identify the current models of CPD and their functions within the context of HE.
- To explore the importance of leadership in higher education and to prevent leadership waste.
- To explore and discuss how reimaging CPD in HE can promote sustainability by introducing transformational and influencing leaders.

2 The Significance of CPD in Higher Education

There are many definitions adapted from the literature for the continuous professional development (CPD); the Organization for Economic Cooperation and Development (OECD) defines the CPD as the activities that enhance and develop the individual learning, skills, knowledge, competence, and expertise along with the expertise of the teacher's characterization mainly reading, reflection abilities, innovative approaches of pedagogical and curriculum development, collaborative learning, and diverse management strategies [11]. Continuous professional development (CPD) is defined as the process of continuous improvement in teachers' efficacy, skills, learning, and competence to increase the student learning outcomes [12]. The additional professional development of the teachers aims to improve the quality of teachers in an informal context as well, including collegial dialogues, personal readings, resource consumption, utilization and conduction of workshops, and peer observations of the in-service teachers [13].

The inclusion of sustainable continuous professional development measures provides efficient teaching leadership and student learning outcomes with a long-term impact [14]. According to the reports of the Center for American Progress, professional learning periodically supports effective teacher-training activities and powerful reformative outcomes, developing a foundation of self-efficacy-based teaching leadership [15]. Moreover, teachers' training in the curriculum, pedagogical innovation, and implementation of evidence-based learning practices can provide better CPD outcomes [13]. Therefore, the landscape of continuous professional development and learning provides long-term sustainable goals for educational outcomes and professional growth as well.

Sustainability is defined as the efficient consumption of resources and generating efficient resilience outcomes [16]. Sustainable continuous professional development is based on the multidimensional learning optimization of the teachers. The CPD's training increases the collaborative learning, efficient knowledge of the pedagogical and assessment criteria, along with the effective peer involvement in the learning activities [17]. The sustainability in the CPD practices will enhance the teacher's participation in curriculum development, and ability to self-analyze and mitigate the change in the organizational culture [12]. It is assumed that the involvement of the teachers in curriculum development activities and pedagogical instruction setup increases collaborative learning skills. However, there is limited research involving the impact of CPD sustainable practices in the curriculum collaborative approaches [18].

According to Bowen [19], the influence of higher education in adult years lasts for approximately 50–60 years. This influence can last for generations in civilization. Higher education systems play an important role in adopting a holistic approach for sustainability: (1) Research and teaching institutes can improve sustainability through development projects and the integration of sustainability practices across areas of study; (2) distinct teachers' practices can impact broader viewpoints through outreach programs; (3) a culture of an organization regarding sustainability aims to raise awareness among the academic staff, as well as regional, and greater populations; and (4) high educational establishments are essential for the onset of next-generation professors. Moreover, by putting in place sustainable campus initiatives such as training and giving rise to leadership roles in HE, it could help in the building of professional and career-oriented students. Universities and colleges may set a good example and encourage their students [20]. Higher education funding is a critical factor in the growth of a sustainable society because organizations with the highest spending on education have strong educational knowledge and professional interest, as well as an intention to embrace and study progressions, such as those pertaining to the SDGs [21].

The value of continuous professional development must not be overlooked since it is a crucial requirement for researchers and practitioners throughout their careers. It can sometimes be prescribed by academic institutions or demanded by standards of ethics or conduct [22]. However, at its foundation, it is the personal obligation of individuals to maintain their knowledge and expertise up to date to provide high-quality performance that protects the community, fulfils customer demand, and meets all the criteria of their profession. CPD leads to increased preservation and standard of living, the atmosphere, stability, ownership, and the industry, based on the area of expertise. This is true in high-risk or highly specialized practice areas, where monitoring on a case-by-case approach is typically difficult. With regard to higher education in the UK, CPD has been the topmost priority and has focused on the development, as well as training of professionals responsible for bringing about a change within the educational system. Within the higher education institutions in the UK, CPD is perceived as a form of "training," "workshop and conferences," or "courses," that are utilized for providing professional development of both teachers, and educators. Ferman [23] suggested a diverse variety of collaborative group opportunities to work with an academic planner, which includes attending conferences, talks among colleagues, exhibiting at conferences, mentoring, and academic research. Those that provide recommendations and suggestions for professional development in higher education realize the significance of such a range of events.

The importance of CPD in higher education with respect to promoting sustainable education lies in the concept of "collaboration." Teamwork is emphasized as a basic pre-requisite in the literature on professional development with regard to higher education. Educators cooperate with their peers on curriculum creation, peer assessment, public and private communication, research inquiry, and other activities. Collaboration can take place inside a faculty, across faculties and professions, between organizations, locally, regionally, and worldwide. Collaboration in teaching and engagement should also be essential components of interaction among educational leaders, and other academic personnel [24]. Therefore, the connection between educational leaders and subordinates is equivalent, and important as the one between professors and their learners [25].

3 Leadership Waste in Higher Education

The leadership in the continuous professional development (CPD) learning produces sustainable educational outcomes by reducing leadership waste, improving learning outcomes and organizational transformation toward the growth of stakeholders. The aspects of quality are an important part of any industry and organization [26]. The quality is analyzed in terms of service delivery and availability of the resources. The educational sector institutions' service delivery at all levels of the primary, secondary, and higher education sector is concretized into different transformational and empowerment aspects of leadership [26].

The concept of "waste" is also important for the service delivery institutions, including HE, which is defined as any human activity that absorbs the resources and provides no value in return [27]. However, the subsequent waste can be removed by addressing the risk factors, challenges, and barriers of the service. The waste is categorized in the literature as "delay," "unnecessary movement," "unclear communication," "errors," "people," "duplication," and "opportunity lost" [27]. The education sector is empowered by internal and external factors. The internal factors are the teachers, principals, and leadership of the institute along with students, and the external factors are the governmental policies and financial resources limiting the role of stakeholders [28].

The case of higher education is complex because of the complex structure, assessment, and diverse expectations of the learning from stakeholders. Therefore, the role of leadership in higher education is important for the effective strategic implementation, policy regulations, and learning incorporation in the classroom as well as in the organizational culture [29]. The concept of leadership arises from the contrary concept of "leadership waste." According to Hartanti et al. [30], leadership waste is the unnecessary activities that do not add value to the learning outcomes of teachers or students. According to the lean theory, the implementation of lean measures in the HE tends to maintain competitiveness among the stakeholders and focus on the market needs, thus improving academic and administrative measures to obtain superior performance of the involved stakeholders [31].

According to Douglas et al. [27], several factors of HE act as the leadership waste in both academic and administrative aspects, including overproduction of material, waste of financial resources, waste of human resources, student dissatisfaction, and poor value to the service delivery and customer preferences of learning and its outcomes. Therefore, reducing the waste generated by high education leadership by increasing efficient service delivery, student satisfaction, and efficient consumption of resources (financial and human) can promote academic service optimization and impact the performance indicators as well [32]. The lean initiatives of the continuous professional development practices can reduce the leadership waste and manpower overconsumption, promote efficient handling of complaints, impact quality learning of teachers and students as well as incorporate efficient research outputs at the universities [31]. Improved efficiency will also free the teaching staff from workload and stress, improving the quality of learning.

The organizational transformation of the stakeholder's involvement in the curriculum and learning practice with CPD can also improve the university outcomes. Several studies introduced that the implementation of lean learning practices with sustainable adaptation at the universities can increase the role of human resources and facilitate the utilization of CPD transformational practices throughout the organization [32]. According to Balzer et al. [26], the implementation of the professional learning in higher education requires first assessment of the institutional capacity, improvement of the capacity, secondly, awareness of the leaders, teachers, and stakeholders in understanding as well as support for lean HE, and thirdly strategic planning for facilitating CPD learning outcomes. Therefore, the processed assessment of the leadership and institutional quality and proposition mitigatory measures with learning practices involving CPD provide a quality transformation of the educational service.

4 Challenges for CPD in Higher Education Within the UK

In terms of professional growth, higher education in the United Kingdom has reached a tipping point [1]. However, previous studies have identified 4 major challenges for HEIs, which include the following:

- The difficulty for HEIs with the establishment of professional standards for higher education faculty would be to ensure that their CPD support is entirely equal, and not confined to already registered professionals. This relates to the demand of establishing a culture in which CPD for teaching is recognized and valued in the same way that CPD for research is, or where continuous professional development is something within which everyone should engage [2].
- The problem for higher education as it establishes a professional conduct structure, and for academic providers who must assist it, will be how to recognize, appreciate, encourage, and allow the documentation, as well as tracking of this diversity of public and private operations.
- CPD includes not just the strengthening teaching skills, but also the assuring of topic information which is current. Exploring the linkages among professional development for education and research is one task for the Higher Education Academy's Subject Centers.

5 Models of CPD in HE

Lieberman [3] categorization distinguishes three forms of CPD: 1. direct instruction that includes training, seminars, conferences, and discussions; 2. school-based learning includes counseling, peer mentoring, action research, analytical friendships, and task-related effective communication leads; 3. non-school-based learning comprises visiting other institutions, education networks, school-university partnerships, and more. Kennedy [4] defined nine CPD models that are highlighted as follows:

5.1 The Training Model of CPD

The CPD training paradigm is worldwide recognized [5] and is, without a doubt, the most popular kind of CPD for educators in recent decades. This CPD method emphasizes a skills-based, technology vision of education, in which CPD enables teachers to renew their talents to demonstrate their professionalism. It is often "delivered" to the teacher by a "professional," with the delivery person setting the aim and the participant being put in a passive role. Although training can be provided within the institution where the individual works, it is most provided off-site and has been frequently criticized for its lack of context to the current classroom scenario within which the team members operate. This approach promotes a high level of centralized control, typically disguised as quality assurance, with a strong emphasis on uniformity and standardization. However, the training methodology has no meaningful influence about how this new information is applied in practice.

5.2 The Award Bearing Model

An award-bearing CPD model is the one that is based on or encourages the completion of award-bearing academic programmed that are often certified by universities. External validation can be viewed as an indication of quality management, but it can also be interpreted as the verification or financing bodies exerting control. As per Kennedy [4], the award bearing model is one of the transmission systems that provide little possibility for instructors to take ownership of their own learning, which is a drawback of this model.

5.3 Deficit Model

This model focuses on resolving inadequacies in an individual instructor. It is often personalized to the person but may be detrimental to the self-assurance of a learner, or teacher, and the building of a communal base of knowledge within the institution. While the deficit model employs CPD for addressing identified inadequacies in education professionals, Rhodes and Beneicke [6] contend that the underlying reasons of low teachers' performance are related not just to teaching staff, but also to leadership, and operational policies. Attributing responsibility to faculty members and viewing CPD as a way of addressing shortcomings proposes a model in which collective action is not regarded, such as the structure itself is not regarded as a potential reason for a teacher's perceived loss to illustrate the intended expertise. It also signals the requirement for a benchmark of ability, and once it is documented, it begins to assert its own power. However, it is argued that individuals are blamed for supposed underperformance in the deficit model, which fails to recognize group accountability within the HE context.

5.4 The Cascade Model

Individual teachers undergo "training sessions," then cascade, or communicate the material to their peers, according to the cascade model. It is frequently used in circumstances where assets are few. Day [7] describes a case study in which a team of instructors used the cascade model for sharing their personal knowledge with coworkers. The group presented its findings, but no serious thought was given to the fundamental values of engagement, cooperation, and accountability that had defined their personal learning. One disadvantage of this paradigm, according to Lieberman [8], is that what is passed down in the cascading phase is frequently focused on skill sets, rarely focused on knowledge, but seldom values focused. Therefore, the cascade model focuses mainly on the knowledge and skills, rather than attitudes and values.

5.5 The Standards-Based Model

As terminology suggests, the standards-based model is based on a focused effective teaching paradigm rather than an effective learning aspect. The standards-based model of CPD ignores teaching as a complicated, situation-specific political, and ethical venture; instead, it "reflects an intention to develop a framework of teachers' education that can produce, and empirically support connections among teachers' efficacy, and student learning outcomes" [9]. It is also greatly dependent on the behaviorist theory of learning, emphasizing the competency of teacher educators as well as resulting incentives at the cost of joint and interdisciplinary learning. Beyer [9] critiques a model for its narrowness and limitedness. There is a lack of focus on key and difficult concerns regarding the goal of teaching, and that "teacher education should be filled with the type of critical examination of societal objectives, future prospects, economic circumstances, and moral orientations."

5.6 The Coaching/Mentoring Model

This model of CPD stresses the importance of one-to-one, or interpersonal communication between two professionals.

Although coaching and mentoring include a one-to-one component, most efforts of differentiating the two claims that coaching is more skill-based, whereas mentoring includes an aspect of "counselling and professional companionship" [10]. While coaching/mentoring approach is defined by its dependence on one-on-one relationships, it may accommodate either a transmission or a transformational view of CPD, based on its philosophical underpinnings.

5.7 The Communities of Practice Model

The question of power is important to effective CPD inside a community of practice. Wenger [11] contends that a community of practice must develop its own concept of partnership, enabling community members to exercise some influence over the initiative. Professional learning should not be regarded as a form of responsibility, or performance management in such a situation. It is asserted that while communities of practice can presumably serve to uncritically propagate power dynamics, in certain circumstances, they can also perform as potent areas of transformation, within which the cumulative effect of individual expertise and experience is greatly enhanced through concerted responsibility.

5.8 The Action Research Model

Proponents of action research, such as Weiner [12] and Burbank & Kauchack [13], say that when it is conveyed through professional learning communities or inquiry, it has a greater impact on practice, yet many populations of training would engage in action research. The action research methodology, nevertheless, does not need the type of teamwork that is seen in a community of practice. According to Burbank and Kauchack [13], collaborative action research offers a solution to the passive state placed on instructors in standard professional development models. They call for instructors to be taught in terms of perceiving research as a systematic process rather than an outcome of someone else's efforts. Sachs [14], on the other hand, relates to the extent to which action research model permits teachers to pose important questions of the socio-political factors that determine the boundaries of their profession.

5.9 The Transformative Model

As per Kennedy [4], this model is one of the most impactful, and transforming models within the context of CPD in education, which provides professional autonomy. Hoban [15] offers an intriguing viewpoint on CPD as a tool of fostering educational transformation. He contrasts the education-focused and culturally vacant model of an instructional approach with the situational-specific approach regarding the communities of practice model, which does not always welcome new kinds of previous knowledge. Such communities use "investigation" as their unifying feature rather than just "practice," implying a far more purposeful, and intentional attitude than is necessary for the case with communities of practice. Transformational model of

CPD is a gist of all previous models discussed and provides power to the teachers so that they can determine and exercise their own learning pathways [16]. However, just like the other models, this model of CPD does not come without tensions either, but since it is the most featured model in academic research, it has the tendency to promote transformative practice within higher education.

Be it direct teaching or training, the conventional concept of CPD is frequently regarded as a top-down delivery model of CPD, in which knowledge on approaches is passed on to instructors for implementation [17]. However, this model has remained quite controversial and critiqued by teachers as it focused on a lecture-style teaching instead of a practical, and active style of learning.

5.10 Reimaging Continuing Professional Development in Higher Education – Toward Sustainability

Students will face new challenges and possibilities in the future. By actively building learners' abilities for problem-recognition and problem-solving, curriculum and teaching methodology may be designed for reflecting the idea that the world will tend to evolve. Problem-posing teaching focuses on students in initiatives, campaigns, and tasks that need research and teamwork [18].

Students must cross-disciplinary barriers to discover feasible, and inventive solutions when confronted with clearly defined objectives. An emphasis on issues and initiatives in learning can assist students with regards to anchoring themselves in their personal observations, help them in perceiving the world as dynamic instead of static, enhance knowledge and judgment, and strengthen students' communication, as well as effective expression skills [19]. However, above all, there is a need of a transformational and effective leader within the context of higher education to promote students' professional abilities, and to bring forth educational reforms. The first step toward a sustainable higher education system is the role of a leader in education who is not confined to a managerial position only but whose knowledge, bureaucracy, and skills are incorporated within the educational institution [20].

The work of Floyd [21] in *"Turning Points': The Personal and Professional Circumstances That Lead Academics to Become Middle Managers,"* sheds some light on the complexities in the role of a Head of Department along with the causes that lead to the HoD playing a middle manager's role. Floyd [21] discussed three contexts that affect the CPD within HE of UK Universities.

1. Changes in the management and culture of a HEI.
2. A change in academic career.
3. An increased complexity in the function and role of HoD.

As per Floyd [21], the key changes that have occurred within the HEIs in the UK affecting the CPD of teachers and students are a high influx of students, and an increase in bureaucracy. There has also been a key change in academic careers as more academics, or educators choose to change their research fields' mid-career so that they can keep up with the evolving research paradigm [22]. However, what seems to be contributing to the evolving nature of HE students and teachers is the role of department head. In the study conducted by Floyd [21], 17 HoDs belonging to different departments were interviewed who belonged to a UK University formed after the year 1992. According to the research, it was found that encountering tension between personal and professional identities exhibited via various socialization experiences across time and may lead to a "tipping point" with a choice that alters a person's career path. However, even though this study could not generalize the findings on a wider population, it served as a milestone for future institutions to understand the role of an HoD, or leader in education to transform the educational framework [23].

It has been suggested that all these developments discussed above might lead to a loss of faith in the discipline, increased responsibilities for academics, deterioration incivility, and a danger to self-identity [24]. Because of these changes in academic careers, as well as enhanced responsibility throughout the sector, the job of educational head of department (HoD) is also evolving and is becoming more complicated as well [20]. Many HoDs appear to be failing in terms of appropriately managing essential components of their work, as research, instruction, and management, as well as personnel concerns, consume an increasing amount of time [13]. According to individual perspective, several colleagues believe that the demands of being an HoD exceed the supposed benefits of the job. While work of the educational HoD is known to be complicated and demanding, some academics appear to like being in this managerial post. These people are referred to as "career-track managers" by Deem [25]. She discovered that certain academicians, especially at post-1992 institutions, would have the desire to intentionally move away from education and research, and perceive taking on a managerial job as a method to accomplish this. They might well be driven by a desire to be viewed as the academic's advocate, assuring that the perspectives of their peers are addressed at the highest levels of the organization [26]. They might become too possessive of their employees, wanting to nurture and support them. There could be many reasons as to why an educationist would take up a managerial position willingly as that of an HoD. However, Floyd [27, 28] found out that, to varying levels, a person's identity corresponds to the organization they represent. The stronger they fit, the more a person is expected to feel a sense of belonging, the more eager they seem to be to take on specific responsibilities, and the more their career options are likely to be altered consequently.

The concept of "Leadership waste" in this context is significant as Leadership waste results from a lack of leadership to reach a higher level of all those working under a leader. It has an impact on all the other manageable problems that the organization faces. If leaders want their organizations to be productive and efficient, they must begin by reducing wastes that they generate [29]. These wastes include structural waste, discipline waste, focus waste, or ownership waste. Within HE institutes, the leaders or HoDs must ensure that there is a comprehensive system within the institute so that focus is maintained throughout the institute for preventing structural waste. This is only possible when educational leaders, or HoDs in this case, are developed from within as per the CPD initiatives such as relational leadership models so that better leadership practice is ensured for promoting improved teaching and research within the HEIs [30].

The relational leadership model works on Knowing-from-within, as well as trying to gain the relevant experience for knowledge and wisdom which is essential, as suggested by [31], and that which relates to adapting ourselves to circumstances demanding skillful discrimination, and skillful reactions. Relational leadership is about building collaborations, engaging with important stakeholders, conversing with peers, and respecting one another in a professional setting [32]. Relational leaders understand the criticality of being aware of the present moment when organizing and solving any dilemma. Stern [33] emphasizes the significance of the present situation and stresses the need for a vibrant conversation between the retrospective and prospective in which present negotiation of meaning must be situationally rooted in the past, but not that much, or the past will regulate us. It focuses on current perception of "interaction" as a relational process of transforming some understanding of our encounter between us in our daily conversations [32], of "having noticed" characteristics and discrepancies within circumstances and helping to bring people into the domain of discussion, all while acknowledging the ethico-moral essence of such connections. Therefore, relational leaders in higher education as part of the CPD promote professional growth, teaching, and research by engaging with co-workers, subordinates, or teachers working under the supervision of an HoD, and research academics [34].

6 Recommendations to Promote CPD in Higher Education

To address these issues, it is crucial that all academics participate in ongoing professional development, which should be seen as an integral part of their jobs in education. This is where continuing professional development (CPD) comes in. As a result, professional development for teachers would be integrated into institutional frameworks and rewarded in the same way that research is. Employees should be assisted in better knowing their individual styles of learning and requirements for taking advantage of existing chances with respect to professional development [35].

There must be acknowledgment, and assistance for the multifaceted character of professional development, which takes place in several educational contexts and involves a wide range of instructional activities. The collaborative aspect of professional development must be increased, enabling for, and fostering contacts among professors inside universities, across professions, and throughout institutions, as well as between all who educate, and facilitate through the aspect of learning [35].

An efficient, professional development has an impact on students, as per [36]. Flecknoe [37] stated that professional development has direct benefits across all levels of educational structure. Mizell [36] went on to say that educators, educational institutions, and community leaders must constantly broaden their understanding and skills to adopt the finest educational methods and to attain the desired results.

Coaching, practice, evaluation, sufficient time, and follow-up assistance are all foundations of successful professional growth [38]. Including the information and abilities earned via professional development programs, the leaders' specific work is integral to their success or failure as leaders in education. Every organization has a unique culture, so it should be handled, and planned properly. There is no fixed strategy for becoming an effective leader that can be learned in a professional development program. Goss [39] stressed the need of developing educational leaders' leadership ability transformation and to enhance teaching education as only competent and professional leadership can secure educational goals. It is proposed that the Directorate of Education provides workshops for teachers on a periodic basis to ensure the strengthening of their abilities in higher education reform. CPD must be explicitly enabled to happen before or after school time, and the Directorate of Education must seek input from institutions' administrators on a routine basis regarding their professional development requirements.

7 Relationship Between CPD, Leadership, and Sustainability

The professional development of teachers is considered an important aspect of sustainable education outcomes in higher education. The sustainable management of the resources of any organization holds immense importance in the overall organizational performance, structure, and growth [40]. As reported by Ceulemans et al. [41], sustainability reports incremental changes in the organization by increasing awareness, engagement of external and internal stakeholders, increasing commutation and reducing the material impact and institutionalization deficiency in the higher education system. Sustainable educational goals are intended to provide personal improvement, enhance decision-making ability, and encourage diverse learning practices [42]. Teachers are the central elements of addressing the educational goals by improving the quality of practice, teaching paradigm, pedagogical measures, assessment criteria, and leadership abilities [17]. The CPD lead professional development is an inclusive measure for embracing peer-oriented activities and collaborative working across different age groups, peer supervision, and collaborative learning [43]. Therefore, sustainability is considered a dynamic tool to plan sustainability changes in the culture of the organization.

The literature lacks research on CPD sustainability to a greater extent. However, some studies have provided a relation between sustainable leadership development in higher education teaching with CPD sustainable measures. The CPD practices affect teachers' abilities, attitudes, and skills and improve teachers' efficacy in relation to the beliefs associated with the education system [17]. Sustainable CPD provides the teachers with an opportunity to reflect on the practices and enact lessons for meeting the requirements of all students through collaborative and individual learning abilities [44]. The CPD sustainable practices also provide an analysis of resource-efficient consumption for the strengthening of educational goals [45]. The sustainable leadership of higher education has diverse knowledge skill, collaborative working ability and management of the resources, finances, and workforce for the benefit of the organization and educational goals [46].

8 Recommendations

Strong organizational and socialization processes within the HEI environment may influence professional development, as well as personal beliefs and identities, which might divert persons who are reluctant or otherwise unable to adjust to such ability to switch identities [47]. The culture of the organization wherein academic studies, as well as the university wherein the department would be centered, provides significant situational knowledge to assist in comprehending the individual's socialization thoughts and feelings, work, and social identities, and, consequently, chosen professional trajectories. Even in the same higher educational institution, departmental cultures appear to differ significantly [27]. As a result, institutions cannot suppose that a standardized approach to Professional development plans would be fruitful. Approaches must be customized to each department, and individual's organization characteristics.

According to studies, educational leaders can create a change in students' educational achievement if they are given the power to make critical decisions [48]. Nevertheless, autonomy alone will not lead to significant improvements unless it is well endorsed. Furthermore, the primary responsibilities of educational leaders must be properly delineated and defined. The obligations of educational leaders such as that of the HoDs should be defined by a knowledge of the practices that will most likely relate to teacher training programs and research as a part of CPD. Therefore, policymakers should promote the following:

Greater amounts of autonomy must be combined with new models of leadership practice, new forms of responsibility, as well as educational leadership training and development [49].

Assisting, assessing, and continuing to develop teaching standards should become a norm in HEIs [50]. Educational leaders must be able to adjust the education program to meet individual needs, encourage collaborative learning, and take part in teacher observing, appraisal, as well as professional development.

Goal setting, evaluation, and transparency should prevail [51]. Decision-makers must focus on ensuring that headteachers and HoDs have the responsibility to establish corporate strategy and that their potential to grow Institutes' aims and strategies and evaluate performance utilizing data to improve practice is maximized.

Collaboration with other HEIs is the new dimension of leadership that must be considered a unique function for educational leaders [40]. This is because it has the potential to help educational systems overall, instead of solely the kids of a particular school. However, educational leaders must hone their expertise to become engaged, and go beyond their immediate sphere of managerial role.

Leadership development is more than just a set of activities or initiatives. It necessitates a mix of formal, and unstructured approaches across all phases and situations of leadership practice [51]. This suggests consistently supporting beginning leadership training, promoting the HE leadership career, organizing training programs, ensuring in-service training covers the need and context, and so on.

Recognize the importance of educational leadership professional associations and provide opportunities, as well as assistance with respect to career growth [52].

These recommendations and discussion of CPD in the context of leadership models and functions of an HoD in reviving and revitalizing the educational context of a HEI serves as a blueprint for future leaders in education so that leadership waste could be prevented within HEIs, and so that a system of teacher education, learning, and research is sustainably developed.

9 Conclusion

From the analysis of various studies conducted in the past and thorough examination of the functions of CPD within the Higher Education context, it can be concluded that HE effective leadership is inherently a complicated issue because it is based on a plethora of interrelated aspects. Several concerns that arose throughout this book chapter indicated topics for additional investigation. However, many healthy insights into the concept of effective leadership within HEIs in the form of HoDs and teachers arose during the writing of this paper. It was found that the value of continuous professional development must not be overlooked since it is a requirement for researchers and practitioners throughout their careers. The importance of CPD in higher education with respect to promoting sustainable education lies in the concept of "collaboration." However, the problem for Higher Education as it establishes a professional conduct structure, and for academic providers who must assist it, will be how to recognize, appreciate, encourage, and allow the documentation, as well as tracking of this diversity regarding public, and private operations. Recommendations to cater to such challenges include providing professional development for all aspects of the academic position (particularly teaching and research), which must be regarded as a routine component of educational staff's professional careers, particularly in the UK. The use of CPD transformational models such as implying action research and transformation model of CPD within HEIs would promote sustainability in terms of improved teaching, and research. Transformational model of CPD is a gist of all other models of CPD, as it provides power to the teachers so that they can determine and exercise their own learning pathways. As per Floyd [27], the key changes that have occurred within the HEIs in the affecting CPD of teachers and students are a high influx of students, and an increase in bureaucracy. Leadership wastes result from a lack of leadership to reach a higher level of all those working under a leader. It has an impact on all the other manageable problems that the organization faces. If leaders want their organizations to be productive and efficient, they must begin by reducing wastes that they generate. There seems to be a need for universities and colleges to better support educators' development as leaders during their careers, as well as to administer the institutional "leadership capability" more efficiently. The findings presented in the chapters largely support the claim that for leaders to be effective, they must be able to involve all relevant parties in their mission in addition to having a clear sense of where they are going and what they hope to achieve. This bolsters the case for creating relational and transformational leadership paradigms in the literature.

Thus, to promote sustainability in education through CPD and incorporating the models discussed, HoDs and other educational leaders associated with HEIs must oblige with the recommendations and models to function as a role model for future educational leaders.

References

1. UNESCO roadmap for implementing the Global Action Programme on Education for Sustainable Development – UNESCO Digital Library. https://unesdoc.unesco.org/ark:/48223/pf0000230514. Accessed 25 June 2022.
2. UNESCO. (2021). *Reimagining our futures together: A new social contract for education*.
3. Nations, U. (2015). The millennium development goals... – Google Scholar. https://scholar.google.com/scholar?hl=en&as_sdt=0%2C5&q=Nations%2C+U.+%282015%29.+The+millennium+development+goals+report.+New+York%3A+United+Nations.&btnG=. Accessed 23 June 2022.
4. UNESCO. (2018). *UNESCO roadmap for implementing the global action programme on education for sustainable development*.
5. Bell, D. V. (2016). *Twenty First Century education:...* – *Google Scholar*. https://scholar.google.com/scholar?hl=en&as_sdt=0%2C5&q=Bell%2C+D.+V.+%282016%29.+Twenty+First+Century+Education%3A+Transformative+Education+for+Sustainability+and+Responsible+Citizenship.+Journal+of+Teacher+Education+for+Sustainability%2C+18%281%29%2C+48-56.&btnG=. Accessed 23 June 2022.
6. Guskey, T. R. (2003). Analyzing lists of the characteristics of effective professional development to promote visionary leadership. *NASSP Bulletin, 87*(637), 4–20. https://doi.org/10.1177/019263650308763702
7. Mourão, L. (2018, September). The role of leadership in the professional development of subordinates. *Leadership*. https://doi.org/10.5772/INTECHOPEN.76056

8. Dyer, G., & Dyer, M. (2017). Strategic leadership for sustainability by higher education: The American College & University Presidents' climate commitment. *Journal of Cleaner Production, 140*, 111–116. https://doi.org/10.1016/j.jclepro.2015.08.077
9. Borko, H., Jacobs, J., & K. K.-I. encyclopedia of, and undefined 2010. (2010). Contemporary approaches to teacher professional development. *researchgate.net*. https://doi.org/10.1016/B978-0-08-044894-7.00654-0
10. Hénard, F., & D. R.-A. I. G. for H. Education, and undefined 2012. (2012). Fostering quality teaching in higher education: Policies and practices. *learningavenue.fr*
11. Marsh, J. A., & Farrell, C. C. (2015, March). How leaders can support teachers with data-driven decision making: A framework for understanding capacity building. *Educational Management Administration and Leadership, 43*(2), 269–289. https://doi.org/10.1177/1741143214537229
12. Roesken-Winter, B., Hoyles, C., & Blömeke, S. (2015, March). Evidence-based CPD: Scaling up sustainable interventions. *ZDM Mathematics Education, 47*(1), 1–12. https://doi.org/10.1007/S11858-015-0682-7
13. J. Y.-J. of T. E. for Sustainability and undefined 2016. (2016). The effect of professional development on teacher efficacy and teachers' self-analysis of their efficacy change. *ERIC, 18*(1). https://doi.org/10.1515/jtes-2016-0007
14. S. S.-R. in S.-E. Systems and undefined 2013. (2010). Learning for resilience, or the resilient learner? Towards a necessary reconciliation in a paradigm of sustainable education. *taylorfrancis.com, 16*(5–6), 511–528. https://doi.org/10.1080/13504622.2010.505427
15. DeMonte, J. (2013). *High-quality professional development for teachers: Supporting teacher training to improve student learning.*
16. Waseem, N., & Kota, S. (2017). Sustainability definitions—An analysis. *Smart Innovation, Systems and Technologies, 66*, 361–371. https://doi.org/10.1007/978-981-10-3521-0_31
17. Ceulemans, K., Lozano, R., & Alonso-Almeida, M. d. M. (2015). Sustainability reporting in higher education: Interconnecting the reporting process and organisational change management for sustainability. *Sustainability (Switzerland), 7*(7), 8881–8903. https://doi.org/10.3390/su7078881
18. Voogt, J., Laferrière, T., Breuleux, A., Itow, R. C., Hickey, D. T., & McKenney, S. (2015, March). Collaborative design as a form of professional development. *Instructional Science, 43*(2), 259–282. https://doi.org/10.1007/S11251-014-9340-7
19. Bowen, H., & Bowen, H. (Eds.). (2018). *Investment in learning:... – Google Scholar.* https://scholar.google.com/scholar?hl=en&as_sdt=0%2C5&q=Bowen%2C+H.+%28Ed.%29.+%282018%29.+Investment+in+learning%3A+The+individual+and+social+value+of+American+higher+education.&btnG=. Accessed 23 June 2022.
20. Littledyke, M., Manolas, E., & Littledyke, R. A. (2013). A systems approach to education for sustainability in higher education. *International Journal of Sustainability in Higher Education, 14*(4), 367–383. https://doi.org/10.1108/IJSHE-01-2012-0011/FULL/HTML
21. Rosati, F., & Faria, L. G. (2019). Addressing the SDGs in sustainability reports: The relationship with institutional factors. *Journal of cleaner production, 215*, 1312–1326.
22. Uçan, S. (2016, December). The role of continuous professional development of teachers in educational change: A literature review. *Harran Education Journal, 1*(1), 36–43. https://doi.org/10.22596/2016.0101.36.43
23. T. F.-T. international journal for academic development and undefined 2002. (2002, November). Academic professional development practice: What lecturers find valuable. *7*(2), 146–158. Taylor & Francis. https://doi.org/10.1080/1360144032000071305
24. S. W.-E. L. S. Centre and undefined 2004. It ain't what you say, it's the way that you say it: an analysis of the language of educational development. *ualresearchonline.arts.ac.uk*
25. Cowan, J. (2001). *Developing skills, abilities or... – Google Scholar.* https://scholar.google.com/scholar?hl=en&as_sdt=0%2C5&q=Cowan%2C+J.+%282001%29.+Developing+Skills%2C+Abilities+or+Capabilities.+Educational+Developments%2C+2%2C+1-4.&btnG=. Accessed 23 June 2022.

26. Balzer, W. K., Brodke, M. H., & Thomas Kizhakethalackal, E. (2015, October 05). Lean higher education: Successes, challenges, and realizing potential. *International Journal of Quality and Reliability Management, 32*(9), 924–933. Emerald Group Holdings Ltd. https://doi.org/10.1108/IJQRM-08-2014-0119
27. Douglas, J. A., Antony, J., & Douglas, A. (2015, October). Waste identification and elimination in HEIs: The role of lean thinking. *International Journal of Quality and Reliability Management, 32*(9), 970–981. https://doi.org/10.1108/IJQRM-10-2014-0160/FULL/HTML
28. Flumerfelt, S., & Banachowski, M. (2011, July). Understanding leadership paradigms for improvement in higher education. *Quality Assurance in Education, 19*(3), 224–247. https://doi.org/10.1108/09684881111158045
29. Bolden, R., Jones, S., Davis, H., & Gentle, P. (2015). *Developing and sustaining shared leadership in higher education.* Accessed 23 June 2022. [Online]. Available: https://academic-cms.prd.the-internal.com/sites/default/files/breaking_news_files/developing_and_sustaining_shared_leadership_in_higher_education.pdf
30. Hartanti, L. P. S., Gunawan, I., Mulyana, Ig. J., & Herwinarso, H. (2022, April). Identification of waste based on lean principles as the way towards sustainability of a higher education institution: A case study from Indonesia. *Sustainability, 14*(7), 4348. https://doi.org/10.3390/su14074348
31. Aij, K. H., & Teunissen, M. (2017). Lean leadership attributes: A systematic review of the literature. *Journal of Health, Organisation and Management, 31*(7–8), 713–729. Emerald Group Publishing Ltd. https://doi.org/10.1108/JHOM-12-2016-0245
32. Klein, L. L., Tonetto, M. S., Avila, L. V., & Moreira, R. (2021). Management of lean waste in a public higher education institution. Journal of Cleaner Production, 286, 125386.
33. Sallis, E. (2014, January). *Total quality management in education: Third edition* (pp. 1–168). https://doi.org/10.4324/9780203417010/TOTAL-QUALITY-MANAGEMENT-EDUCATION-EDWARD-SALLIS
34. Norris, R. (2003). *Implementing the ILTHE Continuing Professional Development Framework.*
35. A. L.-I. and evaluating science education and undefined 1995. Practices that support teacher development: Transforming conceptions of professional learning. *nsf.gov*
36. A. K.-J. of in-service education and undefined 2005. (2005). Models of continuing professional development: A framework for analysis. *31*(2), 235–250. Taylor & Francis. https://doi.org/10.1080/13674580500200277
37. Kelly, P., & M. W.-J. of in-service education, and undefined 2002. (2002, September). Decentralisation of professional development: Teachers' decisions and dilemmas. *28*(3), 409–426. Taylor & Francis. https://doi.org/10.1080/13674580200200189
38. Rhodes, C., & S. B.-J. of in-service education, and undefined 2002. (2002, June). Coaching, mentoring and peer-networking: Challenges for the management of teacher professional development in schools. *28*(2), 297–310. Taylor & Francis. https://doi.org/10.1080/13674580200200184
39. Day, C. (1999). *Developing teachers: The challenges... – Google Scholar.* https://scholar.google.com/scholar?hl=en&as_sdt=0%2C5&q=Day%2C+C.+%281999%29+Developing+Teachers%3A+the+challenges+of+lifelong+learning.+London%3A+Falmer+Press.&btnG=. Accessed 23 June 2022.
40. A. L.-I. and evaluating science education and undefined 1995. Practices that support teacher development: Transforming conceptions of professional learning. *nsf.gov*. Accessed 25 June 2022. [Online]. Available: https://www.nsf.gov/pubs/1995/nsf95162/nsf_ef.pdf#page=58
41. Beyer, L. E. (2002). The politics of standardization: Teacher education in the USA. *Journal of Education for Teaching, 28*(3), 239–245. https://doi.org/10.1080/0260747022000021377
42. Rhodes, C., & S. B.-J. of in-service education, and undefined 2003. (2011, March). Professional development support for poorly performing teachers: Challenges and opportunities for school managers in addressing teacher learning needs. *29*(1), 123–140. Taylor & Francis. https://doi.org/10.1080/13674580300200205
43. Wenger, E. (1999). *Communities of practice: Learning, meaning, and identity.*

44. Weiner, G. (2002). *Professional development, teacher education, action research and social justice: A recent initiative in North Sweden.*
45. Burbank, M. D., & Kauchak, D. (2003). An alternative model for professional development: Investigations into effective collaboration. *Teaching and Teacher Education, 19*(5), 499–514. https://doi.org/10.1016/S0742-051X(03)00048-9
46. Sachs, J. (2003). *The activist teaching profession.*
47. Hoban, G. F. (2002). *Teacher learning for educational... – Google Scholar.* https://scholar.google.com/scholar?hl=en&as_sdt=0%2C5&q=Hoban%2C+G.+F.+%282002%29.+Teacher+learning+for+educational+change%3A+A+systems+thinking+approach.+Professional+Learning.&btnG=. Accessed 23 June 2022.
48. Valentine, J. W., & Prater, M. (2011, March). Instructional, transformational, and managerial leadership and student achievement: High school principals make a difference. *NASSP Bulletin, 95*(1), 5–30. https://doi.org/10.1177/0192636511404062
49. Choi, J., & Kang, W. (2019). Sustainability of cooperative professional development: Focused on teachers' efficacy. *Sustainability (Switzerland), 11*(3). https://doi.org/10.3390/su11030585
50. Deem, R., Hillyard, S., Reed, M., & Reed, M. (2007). *Knowledge, higher education, and the new managerialism: The changing management of UK universities.* Accessed 23 June 2022. [Online]. Available: https://books.google.com/books?hl=en&lr=&id=45cXWj4M0acC&oi=fnd&pg=PR9&dq=Deem,+R.,+Hillyard,+S.,+Reed,+M.,+%26+Reed,+M.+(2007).+Knowledge,+higher+education,+and+the+new+managerialism:+The+changing+management+of+UK+universities.+Oxford+University+Press.&o
51. Floyd, A. (2012, March). 'Turning points': The personal and professional circumstances that lead academics to become middle managers. *Educational Management Administration and Leadership, 40*(2), 272–284. https://doi.org/10.1177/1741143211427980
52. Floyd, A., & Dimmock, C. (2011, August). 'Jugglers', 'copers' and 'strugglers': Academics' perceptions of being a head of department in a post-1992 uk university and how it influences their future careers. *Journal of Higher Education Policy and Management, 33*(4), 387–399. https://doi.org/10.1080/1360080X.2011.585738

Re-designing Higher Education for Mindfulness: Conceptualization and Communication

Damla Aktan and **Melike Demirbağ Kaplan**

In the recent decade, excessive consumption promoted by marketing elicited the highest degree of unsustainable conditions such as climate change, global warming, pollution, and economic scarcity. At the same time, marketing literature offered a myriad of studies focusing on sustainability as a solution to the problem of overconsumption and degradation of world resources. Analyzing the reports published by UN, OECD, European Environment Agency, and other relevant institutions working on sustainability, initiatives taken worldwide seem to be insufficient to save the planet earth. To reveal effective solution to the problem and create a real positive impact for the world, the mindset and behaviors of the world population should be redirected with macro actions.

Both behaviors and mindsets of people should be changed to solve the problem of "overconsumption," through bringing consciousness in thought, referred to as mindful consumption. Although there is an increasing trend in literature to acknowledge the importance and implementation of mindfulness into marketing and consumption practices, both the producers and the consumers still resist to adapt mindfulness into their lives. One of the main reasons of this resistance is not because required mindset is lacking, but instead because the available mindset is not transferred into real action. Generation Y, in particular, who are considered to be the change makers do not display conforming behaviors in terms of mindful consumption considering their mindful mindsets.

D. Aktan (✉)
Independent Researcher, İzmir, Turkey

M. Demirbağ Kaplan
VICTORIA International University of Applied Sciences, Berlin, Germany

© The Author(s), under exclusive license to Springer Nature Switzerland AG 2023
M. Ali S A Al-Maadeed et al. (eds.), *The Sustainable University of the Future*, https://doi.org/10.1007/978-3-031-20186-8_4

1 Mindful Consumption and Mindful Marketing

Global consumption's essential function of serving human basic needs moved beyond its main role [30], and the domain of marketing assumed a key role, by offering "mindful consumption" and "mindful marketing" as important constructs to this aim. Although its acknowledgment within marketing is relatively new, the concept of mindfulness is already discussed in a variety of disciplines, such as social psychology and education [18], quality research [12], as well as reliability subjects in organizational behavior [39], individual and organizational reliability [6], reliability and conflict handling [22]. Several other studies also include research on relationship quality [29], customer orientation [22], innovation and information technologies [35], ideal school and classroom education [9, 28], creativity [25], organizational media uses [37], and past experiences on mindfulness of habitual entrepreneurs [27].

Having a sense of wonder, a feeling of union with nature, a sense of peace of mind, a feeling of wholeness and joy, living in the present movement, and a sense of being accepted within the universe are indicated as the basic aspects of mindfulness [15]. Mindfulness at individual level involves openness to novelty, alertness to distinction, sensitivity to different contexts, awareness of multiple perspectives, and orientation in the present-paying attention to the immediate situation [33]. It is expected that mindfulness requires a crave to overhaul situational awareness on a proceeding premise, to cast question, and test advance to resolve doubtfulness [20], while mindlessness is associated with decreased activation of cognitive, a resulting state of a reliance on past categories like an automatic pilot [18].

Langer [18] states that mindfulness refers to the cognitive qualities of individuals' state of alertness and awareness that is characterized by active information processing, continual creation of new categories and distinctions, explore and attention to multiple perspectives. Brown et al. [5] defines mindfulness as a receptive attention to and awareness of present events, and experience research on mindfulness in business refers to the concept as an ongoing identification of new dimensions of context, which could prove to be helpful in improving foresight and current operations, connecting and sharing of the mindfulness of individuals to create new meaning and knowledge that will help individuals and organizations to achieve greater congruence between their intentions and outcomes [40]. Accordingly, organizational mindfulness includes preoccupation with failure, reluctance to simplify, sensitivity to operations, commitment to resilience, and deference to expertise. Sheth et al. [31] emphasize on the necessity for redirecting the consumption patterns for a more sustainable world with the help of market operations. Mindful consumption is defined as a way to reach this goal. It represents a confluence of mindful mindset and mindful behavior. While mindful mindset is associated with "a sense of caring for self, for community and for nature," the mindful behavior is characterized by "tempering of excesses associated with acquisitive, repetitive and aspirational consumption modes." The customer-centric sustainability is defined as the consumption-mediated impact of marketing actions on environmental, personal, and economic well-being of the consumer [31].

Aside from managerial approaches in literature, marketing has taken a limited focus on mindfulness until very recently. Mindfulness approaches in managerial contexts more focus on individuals' and organizations' ability to achieve reliable performance in changing environments. How they think, how they gather information, how they perceive the world around them, and whether they are able to change their perspective to handle the existing situation are more of the main concern within the managerial field [17, 18]). A study in marketing field was conducted by Malhotra et al. [20] focusing on the mediating role of mindful marketing; however, again not focusing on the consumption side but instead on the effect of mindful marketing on quality orientations, their interaction, and consequences.

More recent studies on mindful consumption include diverse studies such as challenges against transformation to mindful consumption [2], a systematic literature review on the concept [41], views of different consumer segments [21], scale development and validation [13], effects of mindfulness meditation on mindful action and life satisfactions [13], and a critical review of mindfulness and sustainability relationship [36].

Sheth et al. [31] point that attitudes and values of people shape the consumption patterns. A change in both behavior and mindset levels of people through bringing consciousness in thought, referred to as mindful consumption, is deemed as a particular solution to the problem of "overconsumption." The obligation of the humanity to protect the environment regardless of utilitarian concerns is real, and thus, caring for self, caring for the community and the nature is defined as the motivators for behavioral change towards mindful consumption [31]. In this process of behavioral change, marketing is considered to have a potential role in facilitating the mindful consumption, and to advance it by encouraging and reinforcing through the use of product, price, promotion, and place attributes.

Mindful marketing is referred to as an increasingly important notion that aligns marketers' and consumers' interests. The expected mission of such marketing is told to be cultivating mindful consumption through effective, efficient, and ethical ways, while instantaneously considering the interests of both buyers and sellers [32]. Accordingly, marketers should seek ways to find win-win strategies by aligning marketing functions with consumer interests and thus prevent wasteful, unethical marketing.

With all its strategies designed mindfully in order to mind the gap between consumers and marketers, mindful marketing is assumed to lead to mindful consumption, value cocreation, which in return again leads mindful consumption [20]. Mindful consumption-oriented marketing considers about the environmental, personal, and economic well-being of the consumers where consumers control and transform their self-defeating shopping surplus through having a mindset of caring for themselves, the community, and the nature [31].

There may be diverse examples or names given as examples to mindful consumption behavior. In literature, voluntary simplicity, ethical consumption, green consumption, and socially responsible consumption are some other titles that may be under subcategories of mindful consumption or as venues towards mindful behavior. Green consumer, for instance, is defined as "goal-oriented people who take into account the public impacts of their consumptions, aim to create social change, and improve the sustainable development" [38], and behaviors of those consumers are told to be ethically oriented [1]. Peattie [23] moves the focus of research on green consumers from individual consumer to individual purchase. Accordingly, "the degree of compromise" that is the necessity to pay more or travel further and "the degree of confidence" where the consumer knows the environmental benefit or a link to a crucial issue of a product that they buy.

Similar to green consumers, a socially responsible consumer is told to be contributing to sustainability and is defined as "a non-violent person who is in search of spiritual growth, values the beauty of the world, and has an ability to see and to give energy to all the good things in the world" [38]. Socially responsible consumers may also take the form of voluntary simplicity, which is the "singleness of purpose, sincerity, and honesty within, as well as avoidance of exterior clutter, of many possessions irrelevant to the chief purpose of life, an ordering and guiding of our energy and our desires, a partial restraint in some directions in order to secure greater abundance of life in other directions" (Gregg, 1936 quoted in [42]). Taking a twist from Gregg's and several recent authors' formulation of the concept, which also emphasized the spiritual dimension of this life style, voluntary simplicity today is rather summarized as material simplicity (non-consumption-oriented patterns of use), self-determination (desire to assume greater control over personal destiny), ecological awareness (recognition of the interdependency of people and resources), human scale (a desire for smaller-scale institutions and technologies), and personal growth (a desire to explore and develop the inner self) [11, p. 5]. It is defined as the degree to which an individual selects a lifestyle intended to maximize his/her direct control over daily activities and to minimize his/her consumption and dependency [3]. Such examples demonstrate individual behavior patterns that may be categorized under mindful behavior attributes.

2 Mindsets and Mindful Behavior Perceptions of Gen Y

Generation Y, those who were born between 1980 and 2000, is currently between the ages of 18–35. They are largely present today in many of the world countries, while diverse studies may refer to them with slightly different age groups or years [4, 7, 19, 24]. This research focuses on the perception of mindfulness and accompanying consumption behaviors of Generation Y, and how emerging knowledge could be applied to a novel design of higher education curricula. As the data is highly applicable to forthcoming generations which are at their university ages now, the future of the university design would benefit from the findings of this study.

Research presented in this chapter stems from a qualitative inquiry through in-depth interviews, which is conducted on a sample of Gen Y of 40 individuals living in Turkey. Research questions are formulated to explore the presence of a mindful mindset, their definitions of mindful consumption, barriers to mindful behaviors, and several factors that could be of help to contribute to the development of such a mindset. Several dimensions, which are extracted from a corpus of 500 pages of transcriptions, are presented below, followed by a model that depicts the interrelationships between mindsets, behaviors, barriers, and solutions.

3 Mindfulness at Individual Level: Millennium "Tinkerbells"

3.1 Attitude Towards Life: Happy Feet Full of Love and Compassion

While talking about themselves and the concept of mindfulness, instead of addressing money and career first, most of the sample mentioned more about emotions and happiness. They see life as a path or journey that they walk through. They have a perception of seizing the moment, and "happiness" is among the most important concepts for them throughout their journey of life, independent of career or money in most cases. Whatever makes them happy is associated with "good" in life.

> If there is "death" at the end one day, you should be able to say by heart that 'I lived life fully and deeply!'. I only care about being happy. (F, 23)

They care about not being forgotten, and they search for non-material ways of being remembered in the future. Their search for happiness and joy is not only a selfish quest, but also finds a place in their willingness to contribute to life. They are well aware of what expects them in the future, and they see life as a process of learning how to adopt joy into the system they are trapped in. As optimism and diversity is among the most common attributes of this generation's members [43], this can be found in their descriptions of who they are and what they think of life.

> I mean, whatever we do, we should learn doing by joy or learn to become happy with small things in life. We need to integrate different things into our routine lives so that we can get out of that ordinary loop. […]Any contribution I make to environment makes me happy! (M, 24)

3.2 Attitude Towards Self: Uniqueness & Idealism

As mentioned by marketing scholars, diversity and uniqueness is among the core values of this generation [43]. They feel that they are unique and are bored of spontaneous and standardized things brought mainly by industrialization and capitalism. Members of Gen Y in this study support this fact, as their discourses reveal that search for uniqueness and difference is shown up through their instinct of exploring new and diverse things. Sometimes, this search for uniqueness surfaces in their career choices or passions. They mostly attach meaning to whatever they do either professionally or socially.

> Resembling another one, it really disturbs me. Think of consumption world. If something is fashionable in a given year, that year each brand produces the same thing. (F, 23)

Despite this search for difference, simplicity is among the values they are in search of themselves. Yet, they generally feel trapped in the system. They need to adapt themselves to the already set-down rules and regular expectations of the system. They are idealists [26] and try their best until they get what they dream of.

4 Collective Mindfulness: Sharing Is Caring

Looking from collective dimension, it is clear that what they search for and idealistically believe is collective life on earth and thus to serve this ultimate aim.

4.1 Attitude Towards Others: Care Givers, Sharers, and Spiritualists

Generation Y have a desire to have positive impact on people [16], connected to their will to be different and effective in life. They care a lot about helping others even if they have to postpone their own lives for a while. Regardless of age or gender, they become happier when they feel subservient and think that they are the reason for the happiness of another person.

> Living… For me, it means making the people I love feel happy! I mean, some people say 'I die for him/her'. People should not die for someone they care, people should live for the loved ones. (M, 18)

Doing something good for someone else is crucial for them and this makes them feel much better and impactful in life. Any small impact we could create in other lives is perceived as being alive and becoming immortal in the minds of others, and thus decreases the risk of being forgotten.

> …It is so strange, it's as if I am full of beans. I always want to do something good. […] Love and compassion is the most important thing for me. In my social environment, for instance when I can help a stranger, it takes me to a higher level in life. (F, 24)

Members of Gen Y natural sharers. Spiritual life is as important as material life for this generation and sharing is among the important values they care about.

> Attaching meaning to life… It means sharing more. Because life is not only about spending, but instead it is about sharing. It is not about consuming. When you consume. you also consume yourself. (M, 23)

They are spiritually very fragile and inclined to see the good in everything like "Pollyanna" figures, however, when they start to face life and what is brought to the real life by other human beings mostly due to the competitive environment of capitalism, their souls feel somehow threatened and this may result in self-questioning, which leads to the quest for self-evolution through feeling the necessity to change their positive core values. Sometimes, they feel totally lost in the system and try hard for this self-discovery.

4.2 Attitude Towards Environment: Naturalists

Most of the Gen Y, independent of age or gender is very sensitive about world and nature and have high ideals on this issue [10]. While talking about who they are, they frequently mention their belongingness to nature. They further associate the nature with their own identity or body in some cases. Especially, the feeling of collective life is apparent in their sentences.

> Nature and environment is very much important for me. Environmental consciousness means a lot. I mean, when a tree is cut, I feel as if a part of me is cut down. [...]We are in the same life, on same planet, in same place. (M, 23)

This feeling is also valid for other creatures on earth. Being in nature is associated with happiness and collective life for them which should be treated kindly.

> In nature, there are every kind of creature living out there. We, the animals, other creatures… We should be more sincere with the nature so that they can also live there. [...] How good we treat nature and the environment is equal to how good they treat us in return. (F,24)

The nature is seen as an "escape" from the daily routine of the life and as a way of detoxication from the job pressure. Yet, they confirm the perception of continual existence of nature and human-beings as the visitors of this planet, thus nature; and believe that humanity endangers the natural order.

5 Mindfulness on Behavioral Dimension

5.1 Individual Mindfulness: Need Orientation and Logical Thinking Process

Individuality often appears when they talk about personal needs and wants, and in most of the cases, rationality and rational thinking process occur to define the concept in behavioral dimension. Even if they have a positive and high degree of environmental attitude, their basic needs determine their priority throughout their purchase decisions [8]. With slight differences, the findings show that mindful consumption is a term related with the amount human beings "need" to consume for "survival." Need is differentiated from wants and when asked about what mindful behavior means, the respondents mostly focus on personal concerns.

Need assessment and minimization are the two criteria where respondents decide while making their buying decisions.

> How much will this be beneficial for me and how much will it fit to my needs, I first think of this. If I decide 'yes I need this' and assess how long I can use this, sometimes I buy accordingly. (F, 26)
>
> Consuming the minimum of your need, that's what mindful consumption is. Not wasting, not splurging. I am a person who uses something until the moment it becomes unusable. (F, 31)

Mindfulness is perceived as a *logical thinking process* and a state of awareness where there are a lot of information proceeding into the human mind. This logical thinking process is characterized as a *cost-benefit analysis* where they push others to question certain things such as money and the product/service bought in return really replies to the need:

> Buying through cost-benefit analysis... Through thinking whether that product really costs that much and how much it will fit with my needs. (F, 28)

Openness to novelty, alertness to distinction, sensitivity to different contexts, awareness of multiple perspectives, and orientation in the present-paying attention to the immediate situation [33] are among the indicators of mindful mindset, which are visible in the above responses. A Gen Y graduate student defines this buying process as "caution economics" in which she further refers as being alert and cautious while shopping among several messages retrieved from marketing tools.

6 Collective Mindfulness: Balance and Health for All

While Gen Y sample has an individualistic approach, they also balance this attitude with a collectivist approach where they talk about nature, other creatures on earth, and the importance of health for all just as they did in mindset level. Their descriptions do not show a direct parallelization with the mindful behavior definition within the literature whereas they support more the mindful mindset keywords. Yet, they make their own definitions of mindful behavior on a collective level.

6.1 Balance the Nature

Looking back at the literature, the definition of sustainability is more on environmental concerns for this generation [14]. On a parallel tendency, most of the respondents show high level of sensitivity towards nature and environment while defining mindful consumption and explaining about their behaviors.

> Consuming in a way that will both satisfy your need -but your need will be really satisfied; and at the same time, you will not give harm to the environment or anyone else. (F, 23)

In such cases where balancing personal needs while protecting the nature, consumers make the cost-benefit analysis explained in previous section and decide products balancing both and responding both needs.

> I prefer things that are beneficial for me and things that will not harm the nature when I use them… (F, 26)

This feeling of belonging to nature and having a control over it is also a signal of how they feel powerful through having a potential effect on environment through their actions and they act as catalysts for change when considering sustainable consumption [44]. Simple acts and choices of one alternative instead of another one are also observed as a positive contribution to life especially when visual signs are available indicating the care for nature. There are supporting arguments that this generation's priorities are more about environmental-friendly consumption such as acts of recycling.

6.2 Health for All

Most of Gen Y care much about not only their personal health but are sensitive about others as well.

> Buying products that will not be harmful for others. For instance, paying attention to the GDO products, not buying them and preventing others to buy it and thus contributing to the efforts made for removal of these products from the market. (F, 28)
>
> […] I pay attention to by the eco-friendly ones, maybe that's why I spend that much money. I do not use any chemical things to my hair actually but when I do, I try to make natural things such as lemon-honey mixes. (F, 26)

Parents, in the sample, independent of age or gender, also pay attention to what they buy and consume due to health-oriented concerns. Not only in food, but also in clothing and other goods, health orientation is apparent.

> We try to buy natural agricultural products, natural ecological ones, and do not buy products other than such. (F, 27)

Localization is also associated with sincerity in addition to healthiness. While feeling trapped in industrialization, they search for the survival of small local places such as grocery stores, which creates warm feelings, non-artificiality, and humanity under the threat of loss by industrialization.

7 Pro-Consumption Discourses and Barriers to Overcome

7.1 System Pressure vs. Individual Despair

Straughan and Roberts [34] state that unless people believe that they are effective, they are less proactive in buying decisions concerning environment. The most apparent finding among all others that trap the mindful consumption is the feeling of individual despair against the system already established.

> We should protect the nature. As the humanity, we cannot achieve this now, we really devastate the environment. I feel really sorry about this, individually I cannot do anything, there is nothing I can do. I try to pay attention on my own but when I see something bad about nature or environment, I really feel so sad. There is an individual despair I have and the society should become more conscious and mindful to prevent something. (F, 31)

The power of "mindfulness" concept clearly clashes with the presence of mindlessness. The belief of still having a chance or that there is an anti-trust to firms that they really produce products that do not harm the nature are expressed as two main reasons of yet non-organization of society and people.

7.2 Justification

Frequently, despite their top level of sensitivity towards environment both mentally and sentimentally, even though they are aware of the harm they give to nature, other creatures and environment in general by using some of the consumption goods such as plastic bottles, they still continue the same behavior and have several reasons for justifying this pro-consumption behavior. The expensiveness of those consumed goods or services are mostly defined as the excuse for not choosing the alternative.

> Generally, while consuming, I try to consume things that are both healthy and high level of usability. For instance, I try to use glass bottles. But the glass bottles are a bit expensive… (F, 22)

Group psychology and "mob mentality" is another way of justification for mindless consumption. In cases of friendship groups, excess consumption may exist.

> If I am together with friends at the same age, I may sometimes far exceed the consumption of what I need. Instantaneous 'mob psychology' I can say… While in a group, you say 'Let's buy this, let's buy that'; and you see that what you buy far exceeds what you need. (M, 21)

7.3 Non-sacrifice: Power & Happiness Through Shopping

Happiness is among the most frequent reasons of shopping which is told to be outcome of non-sacrifice from shopping, in most cases shows a parallel tendency with justification. Here, the basic distinction lies in the fact that more or less all of the justification is made through the happiness concept. As emotionally motivated individuals, consumption is the solution for unhappiness which prevents sacrifices of those sensitive consumers who seem to care about nature and environment more than anything else. Although nature is of crucial importance for them and they seem to acquire the required criteria of a mindful mindset, they may have subconscious excuses for their contrasted actions or shopping behaviors.

> By shopping, I feel as if I pay back everything, I could not reach even though I really want to. Let's say I argued with a friend, or got a low result from an exam, I just go out for shopping (F, 23)

In a parallel line, justification for non-sacrifice from consumption is associated with therapy which even can be defined through spiritual means and spiritual satisfaction.

> It really makes me happy! I mean, it even changes my aura completely when I am sad. Even if I need it or not, buying something completely changes my aura. It should not be something big, it might even be a small box that I buy to myself. (F, 31)

8 Motivations for Mindful Consumption: "I would Consume Less If…"

The basic answers show that the Gen Y asks for campaigns or ads which will increase their alertness and awareness levels. They express their need for being motivated to decrease their consumption levels through any kind of strategy, which will provide emotional stability and therefore decrease the consumption patterns resulting from unhappiness. They also demand for a system of reproduction instead of consumption.

8.1 Advertising

Gen Y mostly calls for less advertising or a more motivating advertising style that helps them feel a part of nature and belongingness to universe. Even though ads are told to be less effective on their psychology, they still feel the impact of marketing as a pressure over them.

> If there were no ads, we would probably consume less. I would like to see ads that show me the fact that I am also a part of the nature. I mean, the message that when the nature no longer exist, I will no more exist… (F, 23)

A 23-year-old female consumer expresses a similar message where she indicates the expensive cost and heaviness of glass water as an excuse for using plastic instead. Visuality is among the most requested ad type that is expected to have an impact on this generation for behavior change. Aside from visual exposition, simplicity and information about natural world conditions is also defined as another requested attribute in ads.

They are ready to make sacrifices on money or comfort as long as they are motivated and informed to do so. They ask for smart and interesting ads addressing both to emotions and lead people to think, question, and dream a vision at the same time while giving humorist messages. Even though some of them believe that no women can exactly be isolated from shopping, there are points where they still see ads as a way to increase awareness levels, societal consciousness, and thus behavior change.

8.2 Cultural Orientation and Spiritual Self-Development

The cultural background where the consumers grew up mostly determines how they behave throughout their life. Those with parents with mindful consumption patterns have a higher degree of encouragement and willingness to direct more people towards common use of materials.

> This is basically something brought from childhood, very early stages of life. This is totally a consciousness, a mindset given by the family. Family orientation combined with child's inner self orientation; how much s/he has… I mean, this starts in family circle and is related to how much the child improves his/herself. (F, 27)

Respondents declare that emotions are the basic triggers of consumption. Especially, unhappiness shown as the ultimate fuse of spending money is best expected to be overcome by fulfilling the happiness, expectations, and spiritual evolutions of this generation.

Collective action previously requested in ads is further described as an emotional motivator; and in some cases, it is equated with power when compared to individual action.

8.3 Education for All, Education from All Channels

Popularity of social media usage is common, and there are diverse channels of marketing available today. Aside from ads, consumers ask for the use of more than one marketing channel to decrease the consumption levels as only one channel seems to be insufficient.

Necessity to observe the real damage in nature is expressed as another motivating factor for mindfulness expanded through educational seminars. It is more or less requested by all consumers that they need to be informed by any social media or marketing agency in order to be more aware, conscious, and mindful to change their behaviors. Even though the channel might differ, most of the respondents feel the necessity to be addressed directly by marketing channels and indicate that otherwise they can easily ignore.

The model presented in Fig. 1 includes a new conceptualization framework on the topic of "mindful consumption" from the perspective of Gen Y segment in the research. There are two different dimensions and four independently described

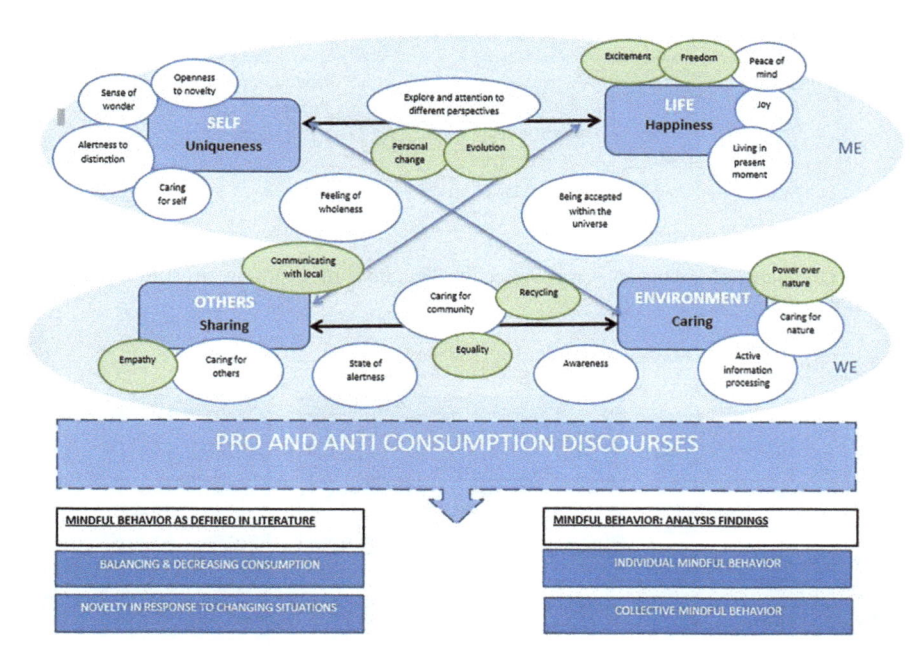

Fig. 1 Conceptualization model of mindful consumption by Gen Y sample

areas in terms of how mindfulness concept is perceived and defined. These two dimensions include a broader level of "individual" versus "collective" orientation whereas the focus areas of these dimensions respectively include "self and life" versus "others and environment."

Each one of these four sub-areas of "self," "life," "others," and "environment" in the model has supporting concepts to mindfulness literature whereas there are also new statements. Both these intersecting concepts and the new ones are shown in the model. Looking back at the literature, there is a gap in the consumption orientations of the sample which complicates the transfer of the mindset to behavioral dimension. This gap is reflected in the model as "pro and anti-consumption discourses" which prevent the practical application of the definition of "mindful behavior" in the literature. Even though Gen Y has pro and anti-consumption discourses, they also have personal discourses about their perceptions of mindful behavior, which are clearly different than the literature. Their discourses also illustrate their resistance to behavioral change while also expressing to what extent they are willing to change their actions, when these are effectively communicated. Figure 2 also shows the groupings of the main emerging themes under two categories of mindset versus behavior segments, while Fig. 3 shows the main emerging themes under diverse categories from the transcriptions.

As can easily be seen from the figure, individual segment is highly consistent with the mindset literature while slightly explains mindful behavior more in terms of collective balance. On the collective side, the focus and care of the generation is more on sustainability for all logic while asks for creating a balance for all the planet.

MINDFUL CONSUMPTION	MINDSET/AFFECTIVE	BEHAVIOR/COGNITIVE
Individual	Happiness Self-Discovery Spiritual Evolution Power	Need Recognition Process of Logical Thinking
Collective	Nature Orientation Balance	Balance

Fig. 2 Groupings of individual versus collective definitions of Gen Y in terms of mindset vs. behavior

Emergent Themes

CHANGE
Personal Change (F,28)
Evolution (F,28)

NATURE ORIENTATION
Power over nature (M,23) (F,19), (F,22)
Belongingness to nature (F,19) (F,27)

SYSTEM
Mechanic Life (F,22) M(23)
Strategic (M,23)
Justice/NonJustice (M,23)
Macro Strategy vs. Individual Despair (F,28) (F,28), (F,33), (F,32), (F,27), (F,22)

CARE
For others (F,28)(F,32)(F,19), (F,30) (F,27)
For Future Generations (F,28)(F,32) (F,30) (F,27)
For earth

BALANCE
Cost/benefit (F,28)
Punishment against harm (F,22)
Obligation and bartering (F,22)
Optimization (M,23)
Balance of Nature: Care for earth (F,27)

SHARING
Empathy (F,28)
Collective Life (F,22)

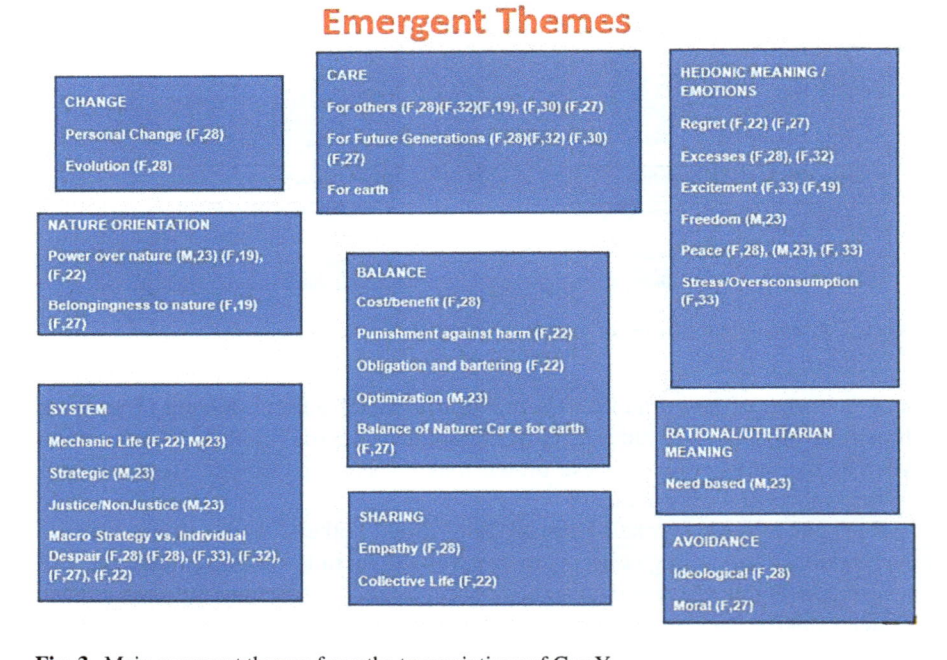

HEDONIC MEANING / EMOTIONS
Regret (F,22) (F,27)
Excesses (F,28), (F,32)
Excitement (F,33) (F,19)
Freedom (M,23)
Peace (F,28), (M,23), (F, 33)
Stress/Overconsumption (F,33)

RATIONAL/UTILITARIAN MEANING
Need based (M,23)

AVOIDANCE
Ideological (F,28)
Moral (F,27)

Fig. 3 Main emergent themes from the transcriptions of Gen Y

9 Application of Findings to New Curricula in Higher Education

9.1 Course Content Formulation for Mindfulness on Behavioral Level

Although many higher education programs already began to offer courses addressing sustainability, corporate social responsibility and mindfulness, the focus of most those courses lies on theoretical concepts instead of behavioral dimension of the topic. As can be easily understood from the quotations of the sample, development of a mindfulness mindset is not enough for changing the behaviors of individuals. There exist barriers against behavioral adaptation of the concept including justification, system pressure versus individual despair, group psychology and "mob mentality," and an unwillingness to sacrifice "power and happiness through shopping." These pro-consumption discourses lead to inefficient behavior formation resulting in lack of mindful action. In order to overcome these barriers, the curricula should incorporate project-oriented contents, behavior-oriented examples of firms, intergovernmental organizations and NGOs, while the assessment should not be based on theoretical exams but instead project evaluations developed through real actions.

Most of the managerial courses include such concepts, while marketing courses focus less on the mindfulness content. Therefore, future curricula of marketing lectures should be more mindfulness oriented as well as increasing the variety of marketing-oriented mindfulness courses. The courses can be diversified into different categories such as advertising for mindfulness, brand creation for mindfulness, marketing strategies for mindful cooperation, marketing for behavior change, barriers marketing professionals have for mindful consumption, digital marketing for mindfulness, among others.

The curricula should include more practical tools rather than delivering theoretical knowledge in order to influence the mindsets of the students and to address the barriers and motivations for mindfulness.

9.2 Mindfulness Centers

Dedicated centers should be incorporated into higher education organization as an administrative department that will specifically focus on the subject. These centers will produce regular projects for students, in which the students voluntarily or professionally work. Additionally, those centers will cooperate with companies and NGOs to figure out the problems of the city and to improve the mindfulness within not only the campus but also the city. Those centers may be managed partially by full-time administrative staff, as well as part-time students, with schedules in accordance to their lecture plans. This will not only give the students a chance to expand their horizons while studying, but will also increase their responsibility through contributing to their mindfulness.

The centers might cooperate with academics in terms of project initiation and application for budget, with students and companies (if needed) for the project application. The centers furthermore could focus on the development of the educational content, which could, for instance, be realized through analog and digital boards around the campus, as well as carrying out concurs for student groups which take the best mindfulness action monthly or periodically.

Finally, the centers may incorporate radio programs to the related clubs of the university, where students will be equipped with necessary information about mindfulness actions, problems, requirements of the campus and the city. Also, those centers could develop advertising campaigns, that is asked by the youth to be fostered for mindful action rather than instinct buying behavior. Those campaigns could be created by the collaboration of media and marketing department students of the campus, which will also enable them to gain insights into ad creation.

9.3 Scholarships for Mindfulness Actions

A strong motivational cause can be scholarships for mindful actions. New kind of scholarships can be offered by universities to motivate students to take collective or individual action for mindfulness. Each mindfulness act can be granted by different amount of scholarship regarding the size of the mindful action taken within the campus.

Furthermore, projects voluntarily realized by higher education students outside the campus with the cooperation of different NGOs or companies shall be granted specific amounts of other scholarships throughout the academic year.

Most universities have systems of certification of achievements for GPA-oriented success. This certification can be expanded to mindfulness actions taken by students as well as lecturers. This would not only enhance motivation for behavior-oriented action, but also increase awareness for group-oriented actions abolishing the barrier of mob mentality and generating positive outcome from group interaction.

The scholarships may be granted by university management as well as business cooperations, which may also foster more collective-oriented mindful action within the country, and also enhancing the university–private sector cooperation.

9.4 Online Digital Boards on Campus

The current Gen Y and the following generations are digital age generations, who are motivated mainly by social media and the "shared information" within all the groups they exist. They therefore ask for training from all sides. To transform their mindsets into real action, they need to become aware of the facts and always be awake and on alert. As most of the time is within the campus for higher education students, they can easily be motivated through keeping their logic always up to date about the mindfulness issues, needs of the world and the potential actions to be taken. They declare that they need to be exposed to necessary information from all the channels in order not to forget to take action or even to motivate themselves to behavior change against the barriers they have.

The university campuses should integrate online digital boards on suitable places where regular informative mindfulness project details may be broadcasted as well as current problems of the city, the country, and the world. Those boards should be controlled by the mindfulness centers, and certain alerts shall be given to the students to remind them each small action they take within a day through lecture times or course breaks. This will not only enable them to be always on alert with an open mind but also give them the chance to collaborate with each other. These online boards may also include tips for spiritual well-being, which is told by the youth as excuses for mindless action when missing. The happier they are, the more they will act mindfully as they will need less justification for their mindless actions such as shopping.

The online boards will also declare the best group of the month who take the best mindful action within the campus periodically. The mindfulness centers will be responsible for this process of selection and transfer of the information to the online boards.

10   Conclusion

Overall, the findings of this research suggest that the Gen Y are "millennium Tinkerbells" who prioritize a happy life full of love and compassion in terms of individual-oriented mindful mindset, while uniqueness and idealism are their core values in terms of self-perception. They are care givers, sharers, and spiritualists, who pay attention to helping others, and balancing life in any way they can. The individual dimension of their mindset, filled by love and compassion, impacts their approach to others and leads them to invest in the well-being other lives, as well as contributing to the balance the nature and environment in any way they can. They search for a meaning in life, in any aspect of life, and talk about meaninglessness of materialism when they perceive the infinity in the universe, the phase of self-evolution throughout life, and the huge potential of undiscovered things waiting out there to be explored. Their spiritually fragile existence also reflects their potential to see the good in everything, while this also leads them to feel lost under the system pressure, just because they reject rules and care more about questioning the reasons of life and existence. They are incorrigible naturalists feeling themselves an essential part of nature and environment. Being in the same life, same planet, same universe with many other living creatures is what determines their attitude towards environment and is certainly associated with happiness. They believe in the power they have to control the nature, to shape the nature, and to help the nature to survive. It would not be wrong to claim that they are missionary spirituals passing through the earth, who feel trapped under the pressure of the capitalist system, and existing rules in this system set by former generations.

Looking at the behavioral dimension, findings of this study conclude that their mindset is not directly transformed into practice. Deviating from the mindful behavior definitions in literature, the model proposed in this study shows that need orientation and a process of logical thinking are the individual dimensions of how Gen Y perceive and define the concept, while balance and health for all are the core of collective dimension. Accordingly, individual mindful consumption requires combining awareness and consciousness with active information processing, cautional economies, need computation and minimization. Of course, what is perceived as "need" by them is a crucial question to be asked, as they also justify their pro-consumption attitudes with the need of happiness. On the collective dimension, their attitude towards others and environment on mindset level shows a similar tendency of how they define the concept. Accordingly, collective mindfulness is balancing and creating health for all living creatures on earth. They attach humanity a mission to balance the nature, consume in a way that will both satisfy the personal needs, by not damaging the environment or any other living creature on earth at the same time. Existing in nature requires controlling the nature on a constructive way rather than a destructive one. By saying health for all, they attach a crucial importance to the fact that none of the consumed goods or services give any harm to personal well-being or the survival of any other creatures and nature's sustainability. It is apparent that despite their high level of awareness on mindset level, there is still insufficiency in behavioral dimension. Systemic pressure and the feeling of individual despair against the system affect how they behave, and these therefore exist as a barrier to mindful consumption. They make comparisons with other European systems, which motivate and foster sustainable practices, and fall into the trap of the illusion of individual despair against the system. They feel as if they are individually insufficient in terms of their potential contribution to change making. This illusion is visible in their discourses such as "What can I do on my own?" Many justifications for their consumption behaviors such as the price and weight of the consumed sustainable products as glass bottles, effects of group psychology or mob mentality, and non-sacrifice due to the power and happiness brought by shopping are also present as barriers to mindful consumption.

Based on the findings, the chapter contributes to the literature through forming a new model of mindful consumption definitions of current generations and as the findings may expand to the following generations, offers suggestions to re-design higher education in order to comply with the needs of the youth to overcome the barriers against mindfulness and to realize behavior change. Alternative suggestions including curriculum re-design, incorporation of mindfulness centers, scholarship formation for mindfulness, and online digital boards are some tools to be used for re-designing higher education for mindfulness. The basic crucial stance here is the way how higher education channels should understand the shift between the generations in terms of understanding of mindfulness as well as consumption habits. Therefore, it is again crucial for the universities worldwide to react in accordance with the changing expectations of the new generation in order to contribute promoting less unsustainable action while also offering solutions for mindful consumption. The essence of change begins with well understanding the dynamics of change makers, who are the Gen Y and the following Gen Z in this stance. The highest impact shall be created by the universities and academia if only the dynamics are well observed and truly directed to create a real impact. Once the pro- and anti-consumption discourses and the quest for more mindful oriented actions from all channels are well analyzed and communicated to the public, it is easier to create the change through the use of marketing and managerial tools to impact the consumption patterns of the future generations as well as encouraging them towards operational actions for sustainability. A real coordination and cooperation is not only necessary but an indispensable part of higher education institutions and policies to understand the current status of the planet and the need versus intentions of the population shaping it both in the past and future. Yet, more suggestions shall be made to further expand the concepts realization and adaptation to higher education in the future.

Acknowledgment This chapter incorporates findings from PhD dissertation and research of Dr. Damla Aktan.

References

1. Anderson, W. T., & Cunningham, W. H. (1972). The socially conscious consumer. *Journal of Marketing, 36*, 23–31.
2. Bahl, et al. (2016). Mindfulness: Its transformative potential for consumer, societal, and environmental well-being. *Journal of Public Policy & Marketing, 35*(2), 1–30.
3. Barton, L. (1981). Voluntary simplicity lifestyles and energy conservation. *Journal of Consumer Research, 8*(3), 243–252.
4. Bakewell, C, & Mitchell, V. W. (2003). Generation Y female consumer decision-making styles. *International Journal of Retail & Distribution Management, 31*(2), 95–106.
5. Brown, K. W., Ryan, R. M., & Crewell, J. D. (2007). Mindfulness: Theoretical foundations and evidence for its salutary effects. *Psychological Inquiry, 18*(4), 211–237.
6. Butler, B. S., & Gray, P. H. (2006). Reliability, mindfulness, and information systems. *MIS Quarterly, 30*(2), 211–224.
7. Chowdhury, T. G., & Coulter, R. A. (2006). Getting a sense of financial security for generation Y. In *American Marketing Association Conference Proceedings, 17*, 191. Chicago.
8. Chu, P-Y., Lin, Y-L., & Chi, W-N. (2013). A study of consumers' willingness to pay for nvironmentally riendly clothing for generation Y: The influences of shopping orientation and green consumption style. *Marketing Review, 10*(1), 19–42.
9. Demick, J. (2000). Toward a mindful psychological science: Theory and application. *Journal of Social Sciences, 56*, 141–159.
10. Duffy, B. (2013). Viewpoint: 'My' Generation: Shared experiences shape individual values and attitudes. *International Journal of Market Research, 55*(4), 2–4.
11. Elgin, D., & Mitchell, A. (1977). Voluntary simplicity. *The Co-Evolution Quarterly, Summer*, 4–18.
12. Fiol, C. M., & O'Connor, E. J. (2003). Walking up! Mindfulness in the Face of Bandwagons. *Academy of Management Review, 28*(1), 54–70.
13. Gupta, S., & Verma, H. (2019). Mindfulness, mindful consumption, and life satisfaction: An experiment with higher education students. *Journal of Research in Higher Education, 12*(3), 456–474.
14. Hume, M. (2010). Compassion Without Action: Examining the Young Consumers Consumption and Attitude to Sustainable Consumption. *Journal of World Business, 45*, 385–394.
15. Jacob, J. C., & Brinkerhoff, M. B. (1999). Mindfullness and subjective well-being in the sustainability movement: A further elaboration of multiple discrepancies theory. *Social Indicators Research, 46*, 341–368.
16. Lancaster, L. C., & Stillman, D. (2002). *When generations collide: Who they are, Why they clash, How to solve the generational puzzle at work*. Harper Business.
17. Langer, E. J. (1977). *The power of mindful learning*. Addison-Wesley.
18. Langer, E. J. (1989). *Mindfulness*. Addison-Wesley.
19. Leschoier, J. (2006). Generation Y… Why Not?. *Rental Product News, 28*, 40–44.

20. Malhotra, N. K., Lee, O. F., & Uslay, C. (2012). Mind the Gap the mediating role of mindful marketing between market and quality orientations, their interaction, and consequences. *International Journal of Quality and Reliability Management, 29*(6), 607–625.
21. Milne, et al. (2020). Mindful consumption: Three consumer segment views. *Australasian Marketing Journal, 28(1), 1–8.*
22. Ndubisi, N. O. (2012). Mindfulness, quality and reliability in small and large firms. *International Journal of Quality and Reliability Management, 29*(6), 600–606.
23. Peattie, K. (1998). Golden Goose or Wild Goose? The hunt for the green consumer. In *Proceedings of the Business Strategy and the Environment Conference, ERP, Shipley*.
24. Pew Research Organization. (2010). Millennials, available at: https://www.pewresearch.org/topics/millennials/
25. Reilly, R. C., et al. (2010). A synthesis of research concerning creative teachers in a canadian context. *Teaching and Teacher Education, 27*(3), 533–542.
26. Reisenwitz, T. H., & Iyer, R. (2009). Differences in Generation X and Generation Y: Implications for the Organization and Marketers. *Marketing Management Journal, 19*(2), 91–103.
27. Rerup, C. (2005). Learning from past experience: Footnotes on mindfulness and habitual entrepreneurship. *Scandinavian Journal of Management, 21*, 451–472.
28. Richhart, R., & Perkins, D. N. (2000). Life in the mindful classroom: Nurturing the disposition of mindfulness. *Journal of Social Issues, 56*(1), 27–47.
29. Saavedra, M. C., Chapman, K. E., & Rogge, R. D. (2010). Clarifying links between attachment and relationship quality: hostile conflict and mindfulness as moderators. *Journal of Family Psychology, 24*(4), 380–390.
30. Shaw, D., & Newholm, T. (2002). Voluntary simplicity and the ethics of consumption. *Psychology & Marketing, 19*(2), 167–185.
31. Sheth, J. N., Sethia, N. K., & Srinivas, S. (2011). Mindful consumption: A customer-centric approach to sustainability. *Journal of the Academy of Marketing Science, 3*(9), 21–39.
32. Sheth, J. N., & Sisodia, R. S. (2006). Does marketing need reform? In J. N. Sheth & R. S. Sisodia (Eds.), *Does marketing need reform: Fresh perspective on the future* (pp. 3–12). M.E.Sharpe.
33. Sternberg, R. J. (2000). Images of mindfulness. *Journal of Social Sciences, 56*(1), 11–26.
34. Straughan, R. D., & Roberts, J. A. (1999). Environmental segmentation alternatives: A look at green consumer behavior in the new millenium. *Journal of Consumer Marketing, 16*(6), 35–55.
35. Swanson, E., & Ramiller, A. (2004). Innovating mindfulness with information technology. *MIS Quarterly, 28*(4), 553–583.
36. Thierman, U. B., & Sheate, W. (2019). The way forward in mindfulness and sustainability: A critical review and research Agenda. *Journal of Cognitive Enhancement, 5*, 118–139.
37. Timmerman, C. (2002). The moderating effect of mindlessness/mindfulness upon media richness and social influence explanations of organizational media use. *Communication Monographs, 69*(2), 111–131.
38. Webster, F. E. (1975). Determining the characteristics of the socially conscious consumer. *Journal of Consumer Research, 2,* 188–196.
39. Weick, K. E., & Sutcliffe, K. (2001). *Managing the unexpected:assuring high performance in an age of complexity*. Jossey-Bass.
40. Weick, K. E., & Sutcliffe, K. M. (2006). Mindfulness and the quality of organization attention. *Organization Science, 17*(4), 514–524.
41. Fischer, D., Stanszus, L., Geiger, S., Grossman, P. and Schrader, U. (2017). "Mindfulness and sustainable consumption: A systematic literature review of research approaches and findings", *Journal of Cleaner Production, 162*, 544–558.
42. Elgin, D. S., and Mitchell, A. (1977). "Voluntary simplicity: Life-style of the future?", The Futurist, 11, 200–206
43. Alch, M.L., (2000), Get Ready for the Net Generation, Training and Development, 54(2)
44. Bentley, M. , Fien, J. and Neil, C. (2004). Sustainable consumption: Young Australians as agents of Change, National Youth Affairs Research Scheme, Canberra, Australia, 1–156

Preparing Future-Fit Leaders for the Sustainable Development Era

Sarmad Khan

1 Introduction

Today's socio-economic growth model is increasingly contested for not fulfilling human aspirations and, perhaps more importantly, accentuating environmental crises and social inequalities [1]. Humankind, through its actions and decisions, has created an imbalance in the ecosystem, which has been paving the way for the complex global challenges we are experiencing today. How do we begin to tackle these challenges and design solutions to advance long-term progress effectively? This question introduces the sustainable development era's (SDE) leadership challenge.

Education is a fundamental human right, a critical global priority, and has a significant role in driving sustainability. Of the 17 Sustainable Development Goals (SDGs), goal four focuses on education and serves as a force multiplier to accelerate essential sustainable development conditions.

This chapter focuses on the specific role of higher education in meaningfully exploring and responding to the SDE's leadership challenge. It discusses the need for higher education to prepare future-fit leaders who can meet the aspirations of the SDGs by creating purpose-driven opportunities and seeking possibilities for a more sustainable way of living and doing—now and into the future. First, I present the rationale behind the necessity for higher education to evolve and better align with the economic, societal, and environmental imperatives of the SDE. Second, I propose a leadership system model that describes how higher education institutions (HEIs) can embrace three core action principles into learner experiences to guide the design of the 'curriculum of the future'. The interaction of these action principles enables learners' adaptation and collaboration that underpins their individual and collective transformation through praxis. The remaining part of the chapter illustrates how the model was integrated into an experiential UN leadership

---

S. Khan (✉)
New York University, New York, NY, USA

programme and the United Arab Emirates University's innovative 'Future of Education' Initiative. The chapter further suggests the model's utility for HEIs to build a 'knowledge to know-how' leadership capabilities approach to prepare students to prosper through change. In the concluding remarks, I summarize the contribution of this chapter.

2 The Sustainable Development Era and the Need to Reorient Higher Education

To understand the context of higher education reorientation in light of the scale of sustainable development frameworks, tools, and models, I will begin with a potted history of the past three and half decades of the field that introduced the SDE.

2.1 The Rise of the Sustainable Development Era

As a construct of Western thinking, modern sustainable development has its origins in the Western ecological movements of the 1960s [2], although ideas about human–nature interactions as part of indigenous knowledge systems carry more longevity throughout their traditional roots. Governments typically saw 'development' as an economic agenda, focusing on conventional macroeconomic parameters.

Since the Stockholm Conference in 1972, where the term 'sustainable development' was formally mentioned, several committees have convened, and manuscripts have been published on sustainable development. In 1987, the ground-breaking Brundtland report titled 'Our Common Future' was prepared by the UN-established Commission on the Environment and Development (WCED), chaired by former Norwegian Prime Minister Gro Harlem Brundtland. The report promulgated a profoundly influential definition of sustainable development and introduced it into the political mainstream. The Brundtland Report defined sustainable development as 'development that meets the needs of the present without compromising the ability of future generations to meet their own needs [3]'. This definition of sustainable development marked the global introduction of conventional sustainable development policy, which placed the environmental debate within the economic and political contexts of international development [2].

The Brundtland Report continues to 'double-click' on sustainable development, opening several frames of insight into the objectives, processes, nuances, interdependencies, pathways, and patterns of sustainable development. For example, sustainable development does not constitute itself around a linear equation, and it does not seek a harmonious fixed state. Instead, it is a process of change where the exploitation of resources, the direction of investments, the orientation of technological development, and institutional change are consistent with the future and present needs [4]. The report further stressed the importance of national, regional, and global collaboration among stakeholders and actors as a precondition to a sustainable future.

The Brundtland Report provided the ground stone for convening the 1992 Earth Summit in Rio de Janeiro. The Earth Summit itself represented a significant step forward in the sustainable development discourse, with outcomes such as the Framework Convention on Climate Change, the Convention on Biological Diversity, Principles of Forest Management, the Rio Declaration on Environment and Development, and notably Agenda 21. The 600-page Agenda 21 is a non-binding sustainable development action plan of the United Nations to put sustainable development principles into practice [5].

Since the Brundtland report and the first Rio Summit, sustainable development has become a desirable goal for which implementation has proven to be complicated. In his 2002 report on implementing Agenda 21, United Nations Secretary-General Kofi Annan raised the undoubted gap in its enactment, noting that progress toward reaching the goals set at Rio was slower than anticipated [6]. Chabrak and Richard [7] explained that the main reason behind this gap might be the difficulty of conceiving the move from theory to practice, which the elusive definition of sustainable development could partly explain. Although sustainable development has gained currency within governments, NGOs, prominent international organizations, and the private sector, the flexibility of the sustainable development concept has enabled many actors to adapt it to their purposes. Its suppleness has consequently led to various interpretations and confusion that has compromised the effectiveness of its implementation. Several voluntary initiatives did, however, emerge, including the World Business Council on Sustainable Development, the OECD Round Table on Sustainable Development, United Nations Global Compact, Caux Round Table, Equator Principles, Global Reporting Initiative, Young Global Leaders, Global Business Oath, Extractive Industries Transparency Initiative, CERES principles, and Business Leaders Initiative on Human Rights. These initiatives and many others prove an encouraging trend but not necessarily a real drive for sustainability.

The financial sector also plays a role in the slow progress towards sustainability. Achieving the SDGs requires action within the real economy and the financial industry. Although the global economy has abundant stocks of financial assets, investment flows for long-term sustainable development are insufficient, as highlighted by the UN Environment Programme [8]. Between $3.3 and 4.5 trillion per year needs to be mobilized to achieve the 2030 Agenda. While an objectively prominent figure, this amount constitutes just a tiny fraction of the $87 trillion in gross world output, according to the International Monetary Fund [8]. In his 2021 report 'Our Common Agenda', the UN Secretary-General highlighted the need to go into emergency mode to reform global finance [17].

Finally, a critical driver of sustainable development is education. Of import to the realm of education, Agenda 21 highlighted the potential of the scientific and technological communities to make meaningful contributions to policies concerning development and the environment. It emphasized the role of academia in that effort [9]. The initial thoughts concerning Education for Sustainable Development (ESD) were captured in Chapter 36 of Agenda 21, focusing on three key programme activity areas: reorienting education towards sustainable development, increasing public awareness, and promoting training [10]. This chapter discusses how the education sector and its HEIs could play an essential role in accelerating progress towards achieving the SDGs.

2.2 Navigating the Turn into Higher Education for Sustainable Development

Over a decade after the Earth Summit, the United Nations declared 2005–2014 the Decade of Education for Sustainable Development (DESD), and HEIs were recognized as significant contributors to the advancement of sustainability. It was throughout this period the concept of ESD matured through comprehensive implementation measures and action plans that by and large emerged from Agenda 21 and further matured. The particular role of HEIs in the matter is articulated below:

> Countries could support university and other tertiary activities and environmental and development education networks. Cross-disciplinary courses could be made available to all students. Existing regional networks and activities and national university actions which promote research and common teaching approaches on sustainable development should be built upon, and new partnerships and bridges created with the business and other independent sectors, as well as with all countries for technology, know-how, and knowledge exchange. [10]

As the lead United Nations agency for ESD and responsible for the global management, coordination, and implementation of the global ESD 2030 strategy, the United Nations Education and Scientific Organization (UNESCO) offers up a few workable definitions:

> Education for Sustainable Development (ESD) empowers learners with knowledge, skills, values, and attitudes to make informed decisions and take responsible actions for environmental integrity, economic viability, and a just society. [ESD] is a lifelong learning process and an integral part of quality education. It enhances the cognitive, social and emotional, and behavioural dimensions of learning. It is holistic and transformational and encompasses learning content and outcomes, pedagogy, and the learning environment itself. ESD is recognized as a key enabler of all Sustainable Development Goals and achieves its purpose by transforming society. [11]

Although several definitions of ESD have been formulated, there is not a single interpretation and use that has been universally approved. The term will continue to evolve, as many processes do. For contextual purposes of this chapter, the definition provided by Wals [12] is most adequate and mainly congruent with UNESCO's articulations:

> Sustainable development education is a learning process (or a teaching/training approach) based on the ideals and principles that underlie sustainability and is concerned with all levels and types of education. [12]

As the UN DESD concluded in 2014, including the end of the Millennium Development Goals (MDGs), global consultations were well underway to prepare a new global development framework with a new timeline. In 2015, the 2030 Agenda and its 17 SDGs were adopted by all 193 United Nations Member States. The 2030 Agenda offered a vision that balanced previous assertions of the economic focus of development with one for socially inclusive and environmentally sustainable economic growth. Of the 17 SDGs, SDG4 is dedicated to education. The matter of access to higher education is addressed in target 4.3 of SDG4, which aims to ensure all women and men have equal access to affordable and quality technical, vocational, and tertiary education, including university [13]. However, target 4.7 of SDG 4 firmly recognizes the importance of developing capabilities and calls for all learners to acquire the knowledge and skills needed to promote sustainable development [13].

With 'universality' as a defining feature of the 2030 Agenda, the SDGs expanded the education focus beyond primary and secondary levels to include tertiary education. This was an essential shift as higher education was missing from the international development agenda, evidenced by the previous set of eight MDGs and the Education for All framework. For example, the MDGs focused on progress within discrete sectors like education and health. By contrast, the SDGs build on the MDGs but link them to more expansive, economic, societal, and environmental systems to achieve the transformation to sustainable development [14]. The SDGs have also set up second-generation leadership challenges that all actors are individually and collectively involved in and must grapple with. These are not technical challenges that can be solved by adjusting or improving current practices or policies. They require sustained global intent and attention at the highest levels of leadership and influence.

Since the Bruntland Report, and particularly over the last few years, multilateral and plurilateral organizations and forums have recognized the fundamental role of education in creating healthy and inclusive societies, equipping people with future work skills, and cultivating pathways for students to better transform their resources into achievements they have reason to value. The 2018 G20 Education Ministers' Declaration [15] recognized education as a driver for sustainable development and called for promoting multiple and flexible pathways into lifelong education and training, and education that keeps in step with technological innovations. The World Economic Forum [16] further stressed the importance of building future-ready education systems and curricula built for the twenty-first century. The United Nations 'Our Common Agenda' [17] perch education and skills development as a high priority in building capacities to help people navigate several transitions throughout their lives. The International Commission on the Futures of Education, established under UNESCO, leaned the furthest to propose a new social contract for education, grounded in the principles of the right to education, and a commitment to education as a public societal endeavour and a common good. The Commission's report unravels the dimensions of the social contract and how conventional ways of thinking about education, knowledge, and learning deter movement towards the desired future [18].

The corollary: the most significant contribution of an HEI to sustainability and ultimately to society is the pursuit of ESD; however, simply expanding existing growth and educational development models is an insufficient way forward. Suppose conventional thinking about education, knowledge, and learning indeed deter movement towards the desired future. What new thinking is needed to rewire learning to bring the future of higher education into the present?

3 The University of the Future in the Sustainable Development Era

This section discusses what a 'university of the future' could embrace to graduate future-fit leaders in the SDE. First, I briefly discuss the characteristic nature of sustainable development problems and the role HEI can play in addressing them. Second, I propose a regenerative leadership system model for sustainable development called the 'ACT Model'. I explain how HEIs can adopt this model to enable students to gain adaptative and collaborative capabilities that underpin their individual and collective transformations to effectively deal with the complex challenges of the SDE. Third, I explain the three action principles of the ACT model that universities need to incorporate into learner experiences as prerequisite measures. These action principles foster capabilities development, knowledge curation, and networked collaboration. Lastly, I present implementation case examples of the ACT model in an innovative UN leadership programme and a university, reflect upon this experience, and discuss its utility for HEIs in pursuit of ESD.

3.1 The Difficulties Will Argue for Themselves: From Complex Problems to Sustainable Solutions

As explained in the previous section, the sustainable development path is fluid, dynamic, and ultimately uncertain. The leadership challenges of the 2030 Agenda and its SDGs are not technical and cannot be solved by improving current practices and policies. Politically, they demand sustained responsiveness at the highest levels of global leadership and influence. Practically, they require leaders to navigate complex problems and press a deeper questioning of the fundamental ways of how we collaborate, search for patterns, and explore and test ideas in a constantly changing world.

Complex or 'wicked' problems rely on a high degree of interdependence to be solved. These challenges are fluid, often unpredictable, and don't lend themselves to one-size-fits-all solutions. Instead, they generally require a range of potential solutions to be considered and combined through rapid iteration and testing. The Cynefin Framework (Fig. 1), developed by Dave Snowden in 1999, is a decision-making tool that provides a useful method to lead in uncertainty. It suggests five

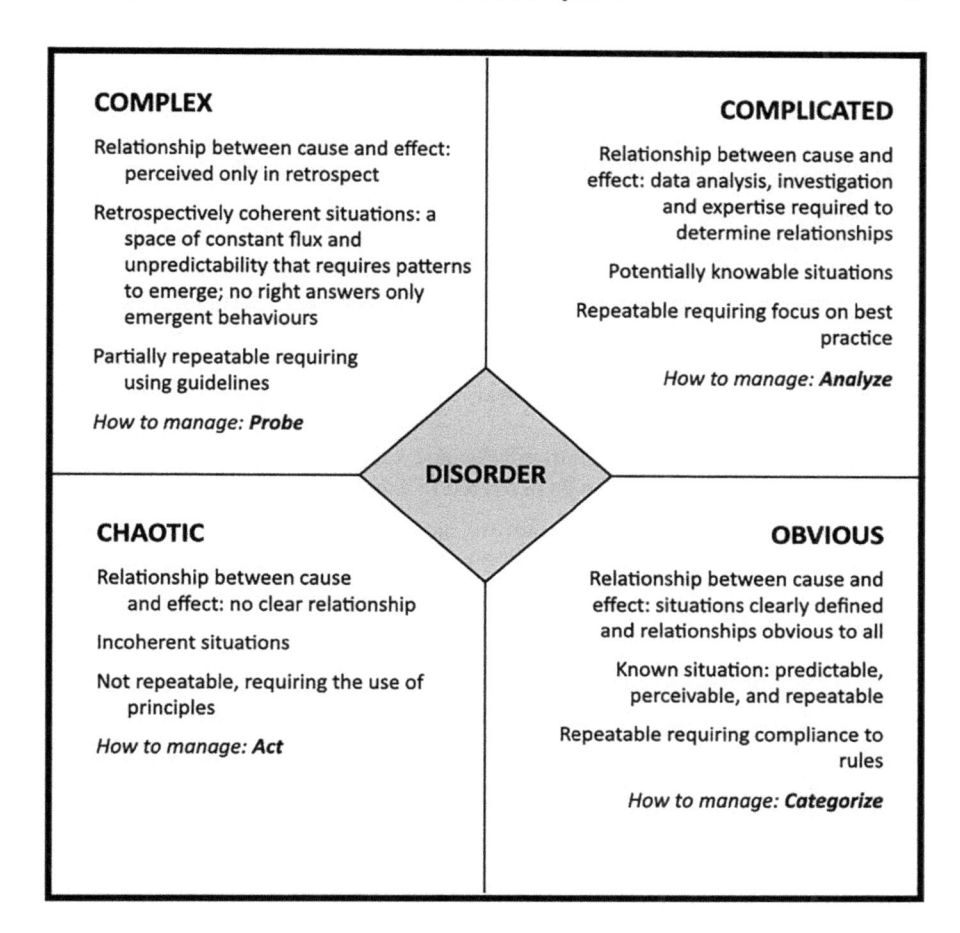

COMPLEX	COMPLICATED
Relationship between cause and effect: perceived only in retrospect	Relationship between cause and effect: data analysis, investigation and expertise required to determine relationships
Retrospectively coherent situations: a space of constant flux and unpredictability that requires patterns to emerge; no right answers only emergent behaviours	Potentially knowable situations
Partially repeatable requiring using guidelines	Repeatable requiring focus on best practice
How to manage: **Probe**	How to manage: **Analyze**
CHAOTIC	**OBVIOUS**
Relationship between cause and effect: no clear relationship	Relationship between cause and effect: situations clearly defined and relationships obvious to all
Incoherent situations	Known situation: predictable, perceivable, and repeatable
Not repeatable, requiring the use of principles	Repeatable requiring compliance to rules
How to manage: **Act**	How to manage: **Categorize**

(Center: **DISORDER**)

Fig. 1 The Cynefin framework [19]

domains that categorize problems: obvious, complicated, complex, chaotic, and disorder (if the problem has not determined) [19].

Beginning with the 'disorder' domain, the existence of a problem is known; however, the source is unknown. The efforts would need to be made to seek information and patterns to move the problem into one of the other domains [20]. While obvious or 'simple' problems are discrete problems that might not be easy to address, the solution and its impact are generally known, at least in theory, based on good practices and lessons learned elsewhere. For example, addressing water pollution at the city level can be traced and handled with good practices or targeted programmes. 'Complicated' domain problems are different as their nature and consequences are not visible immediately and require deeper investigation and the involvement of subject matter experts. For example, fighting against corruption, instituting tax reforms, establishing trade agreements, and other such examples represent the nature of complicated problems. 'Chaotic' problems are in the realm of the unknowable. They are in deep turbulence with no clear cause-and-effect relationships, which makes searching for them futile. Chaotic problems can be addressed through direct command and control mechanisms while attempting to seek different points of view to bring the problem back into the realm of the complex [20].

The framework defines 'complex' problems as typically unpredictable situations that require time to see instructive patterns of cause–effect relationships emerge [20]. These are the types of problems we face in pursuing sustainable development. For example, solutions to integrated and interdependent challenges—such as the SDGs—can best be sought through collaborative experimentation, allowing competing emergent ideas to be developed and tested by involving diverse teams and partners. To illustrate the nature of complex problems and the changes HEIs need to implement to help tackle them, I draw on a historical case known as the 'Mulberry Piers'.

On May 30, 1942, Winston Churchill dictated a short memo with the subject title 'Piers for Use on the Beaches' to be delivered to Admiral Mountbatten [21]. It read:

> They must float up and down with the tide. The anchor problem must be mastered. Let me have the best solution worked out. Don't argue the matter; the difficulties will argue for themselves. —Winston Churchill

On the morning of June 6, 1944, precisely 2 years shy 1 week from the date of the memo, World War 2 Allied Forces (US, UK, Canada) stormed the beaches of Normandy in Nazi-occupied France, what is historically known as D-Day. Once the troops were there, food, oil, military equipment, and medical supplies needed to continuously arrive from the UK and support the war effort. To do so, the British had to build two portable harbours, transport them across the English Channel, and assemble them off the coast of Normandy under fire from enemy guns. And they did. Later that D-Day afternoon, disassembled sections of two harbours weighing 1.5 million tons were towed by 170 tugboats that departed the UK and set sail for France [22].

Among several other supporting structures, the harbours required 16 km of roadway that had to twist on supporting pontoons and mile-long piers code-named 'Mulberry Piers' [23]—for perspective, each pier was twice the length of the Burj Khalifa, the current tallest building in the world. The Mulberry Piers was the most outstanding engineering achievement of the time resulting in ground-breaking technology involving the combined efforts of civil and military engineers and construction teams over 2 years. It required rapid prototyping and extensive testing—in top secrecy—to develop the best solution to a problem. This call for innovative and solution-driven thinking, particularly in matters of urgency, is as essential now as it was then.

Taking a lesson from the Mulberry Piers, implementing the integrated and universal nature of the SDGs constitutes a complex problem. The best solutions need to frame the problem holistically by recognizing the interdependency of its parts and involving multiple diverse actors. The global community recognized that the previous ways of thinking and acting are insufficient and require examining what to conserve and discard from past practices and inventing new ways to build and adapt.

We only need to browse our memory back to 1 year ago to draw an important example of global collaboration on an unprecedented scale involving governments, medical practitioners, researchers, vaccine companies, tech companies, philanthropists, foundations, businesses—and importantly, citizens—to respond to the COVID-19 pandemic. The unprecedented impacts and consequential social, economic, and humanitarian needs introduced or exacerbated by COVID-19 drove the rapid development of vaccines and the adoption of new digital technologies at scale. These partnerships were harnessed to support the public health and education response worldwide. The trajectory of generating digital solutions was accelerated by greater innovation, investment, and access across several levels.

The space for finding solutions to complex sustainable development problems requires interacting with all parts and levels of society well beyond the traditional focus on national and local government and structures, primarily through the active involvement of industry and academia, and mainly its HEIs. This entails concentrating HEI's focus to enabling students to analyze, differentiate, define, and respond to these increasingly complex problems and collaborate and co-create sustainable solutions. But how could this be translated into the new curriculum? What measures need to be adopted to support ESD?

In the following subsection, I introduce the 'ACT Model' as a framework to guide HEI curriculum reform to support ESD. This model depicts a regenerative leadership system, which enables adaptation and collaboration that underpin individual and collective transformations. It promotes a new 'software' that infuses learner curiosity, experimentation, and pathways that help adapt mindsets and behaviours to drive collaboration and demonstrate the 'theory to praxis' continuum of transformational change.

3.2 The ACT Model: A Regenerative Leadership System to Support Sustainable Development

During my time as the Head of Leadership Development at the United Nations Development Operations Coordination Office (UN DOCO) from 2014 to 2019, I, along with my colleague, Ifoda Abdurazakova, embarked on extensive consultations with leading experts and academics in the field of leadership, systems theory, and organizational design at MIT and Harvard University, including meaningful discussions with Otto Scharmer and Jorrit de Jong on our emerging thinking, ideas, and approaches. Drawing on compounded analysis and insights of delivering leadership programmes and a deep understanding of the unique role and complex multidimensional country contexts in that UN leaders operate the components of the ACT model were conceptualized (Fig. 2). These components became the new wireframe applied to design new and re-design existing leadership development programmes for country representatives of the Secretary-General (called 'UN Resident Coordinators') and UN leadership teams to navigate better the complex sustainable development terrain with national partners [24].

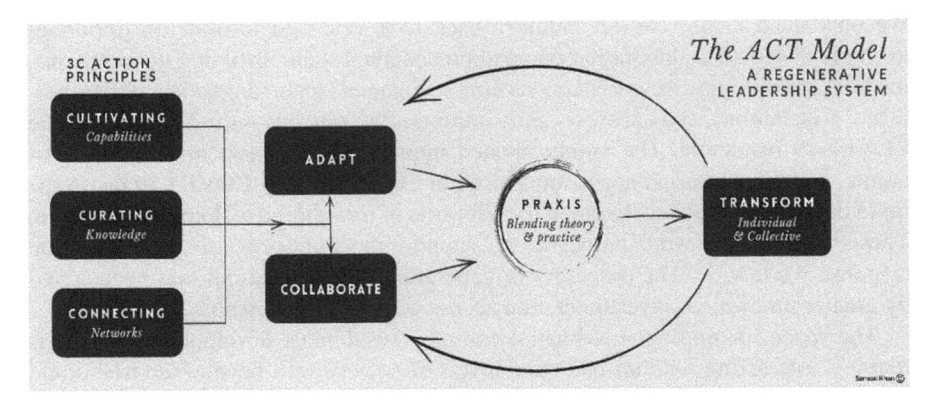

Fig. 2 ACT model: a regenerative leadership system

Before presenting how the ACT model was integrated into the UN leadership development programme, and another HEI initiative, I first discuss its components and foundation—the 3C action principles.

3.2.1 The Components of the ACT Model

Adapt

To unpack sustainable development challenges, adaptive thinking and behaviours involve rethinking complex, multidimensional problems and constructing tailored solutions based on the knowledge of issues and local contexts that enhance individual and team leadership, particularly in larger systems with constantly changing environments. Interplacing diverse perspectives and interests in local and regional settings with global implications requires individuals to handle tensions, dilemmas, and trade-offs supported by systems thinking. For example, balancing equity and freedom, autonomy and community, innovation and continuity, efficiency and the democratic process, and striking a balance between competing demands rarely lead to an either/or choice or even a single solution. Individuals need to think in a more integrated way that avoids premature conclusions and recognizes interconnections, expanding knowledge and boosting intellectual capital as new information becomes available. Individual stewardship through adaptive approaches in response to changes or fluctuations present in a person's external environment, including co-creative endeavours, can allow meaningful collaboration to flourish. In a world of interdependency and conflict, people successfully secure their well-being and that of their families and their communities only by developing the capacity to understand the needs and desires of others, considering the interconnections and interrelations between contradictory or incompatible ideas, logics, and positions, from both short- and long-term perspectives.

Collaborate

Long-range positive impact and scalable solutions require collective 'as one' approaches to building a shared understanding of problems, enabling joined-up support for effective solutions beyond the scope of individuals and organizations. It involves connecting networks and solving collective action problems through new ways of multilevel coordination to confront evolving contextual realities and desired outcomes. This collaboration requires 'agile coordination' between flexible and adaptive networks that can respond to changing trends and new information, continuously improve approaches to deliver more value outcomes, and test and iterate new ideas, models, and solutions. Agile coordination requires networks that can rapidly interact and collaborate with key system components, such as other HEIs, students, private and public sector entities, and local, regional, and global communities. This lattice of engagement would be enabled by curating knowledge and connecting networks prerequisites supported by digital technologies that will constitute an effective sensing and responding system.

Transform

To bring about long-range positive impact and scalable solutions, the transformation of how individuals and societies address problems is needed to better those societies. The benefit of collaboration is to enable shared resources and capabilities to bring greater social and economic value and impact. To understand this dimension in the model, we need to consider two aspects: (i) the relationship between individual and collective (societal) transformation and (ii) the importance of praxis to move individuals and society from the level of adaptation and collaboration to a transformation stage. According to Giddens' structuration theory [25], an ontological framework developed for studying human social activities in terms of the relation between the natural and social sciences and the connection between the individual and society, human social action is influenced by the 'duality of structure'. In other words, social structures are both constituted by human agency and yet, at the same time, are the very medium of this constitution (the condition). This means that incommensurable forces of social structures do not constrain human agency, but at the same time, it is not simply a function of the individual expression of will. An individual's social actions are a synthesis of the effects of agency and structure—as structures influence actions, and the structures themselves are socially constructed, maintained, and adapted through the exercise of agency. Therefore, social actors are engaged in producing and reproducing their social world. People shape society but with resources and 'practices' inherited from the past. Since most of our practices occur at the level of practical consciousness, which is generally informed by 'mutual knowledge,' that is, knowledge taken for granted and based around 'rules' and 'habitus' [26], people reproduce their social world and rules.

Since the agents and their practices reproduce structures, depending on circumstances, adaptation and collaboration become the only means to cease prevailing thinking and behaviours that impede progress on achieving the SDGs and allow for individual and societal transformation. This brings attention to explaining the critical role of praxis in the transformation process. The traditional Aristotelian, Platonist, and Cartesian dualisms split mind from body and theory from practice. It elevated the mind (theory) to become confined to the elite, while any form of practical activity is seen as 'base' and is carried out by the 'average man'. For Hannah Arendt [27], Western philosophy has wrongly focused on the contemplative life at the expense of the active life. This duality prevents humans from realizing full humanity, as it can only eventuate when the mind and body are working in unity. Praxis seeks to overcome this duality that sees humans divested of either their theoretical or practical capacities by re-establishing the importance of reflective human activity that transforms the natural and social world. For Gramsci [28], 'the philosophy of praxis does not tend to leave the simple in their primitive philosophy of common sense but rather to lead them to a higher conception of life'. The phrase itself shows how inseparable thought from action, theory from practice, and philosophy from revolution are. By combining reflection and action, praxis becomes the condition of individual and societal transformation enabled by continuous adaptation and collaboration.

3.2.2 The 3C Action Principles: ACT Model Foundation

The components of the ACT model discussed above offer HEIs a regenerative systems framework that would prime learner preparedness for the complex sustainable development world and evolving labour markets. To enable students to adapt, collaborate, and transform, HEIs need to place solving the world's most pressing challenges at the centre of their mission so that the complex problems of the SDE become the driving force for what and how students learn. For that purpose, HEIs need to implement the 3C action principles as a foundation to guide future curriculum design and structure.

First, HEIs must re-orient focus on equipping next-generation leaders with the capabilities required to navigate change and complexity. Second, HEIs should promote interdisciplinary and transdisciplinary curriculum development and systems thinking approaches that link diverse expertise, knowledge, and ideas, and give rise to potential integrative disciplines. For instance, HEIs need to foster collective actions and knowledge mobilization and curation across all segments of society, which requires bringing individuals to work together for a common cause, despite often being led by different motivations. Third, HEIs need to infuse learning experiences with a participative process to problem definition and concerted action. Because viable solutions exceed the capacity of any one actor on its own and require understanding the changing dynamics at the global, regional, and country levels, HEIs need to become more entrepreneurial and innovative in creating joined-up opportunities to cultivate prospects for students to collaborate and co-create with actors from the quadruple helix (academia, government, industry, and community).

Cultivating Capabilities

The 'curriculum of the future' needs to cultivate students' capabilities to nimbly manoeuvre complex spaces, lead sustainable development efforts and think with a systems mindset seeking solutions to existing and emerging complex sustainability problems. Therefore, HEIs are called to shift the focus from hard to soft or 'durable' skills. Because hard skills—those professional or technical skills that are often job-specific—are now becoming obsolete more rapidly, often within just a few years, the soft skills and competencies that contribute to adaptability, interaction, and resilience are much more valued. According to the WEF report on the 'Future of Jobs' [29], if previous industrial revolutions did not require a rapid change in building the training systems and labour market institutions needed to develop significant new skill sets on a large scale, with the upcoming pace and scale of disruption brought about by the Fourth Industrial Revolution, this may not be an option. Current technological trends make nearly 50% of subject knowledge acquired during the first year of a 4-year technical degree outdated by the time students graduate. For example, the rising computing power makes working with data and making data-based decisions an increasingly vital skill [29].

Soft skills are durable because they are transferrable across jobs and careers with greater lifelong relevance. For these reasons, soft skills become the core skill set for the future of work and a driver for a systems leadership approach needed to prosper in the SDE. They indicate that an individual can adapt and learn as the world changes. The distinctive competencies required here include foundational skills like academic and cognitive abilities acquired through formal education, work experience, training, and other competencies gained in informal ways. They are crucial for learners to be able to operate in a complex, uncertain, and ever-changing world, to handle non-routine and abstract work processes, to have the ability to make decisions, and to have the requisite understanding to handle system-based and interactive work. HEIs should infuse in learners' capabilities for reasoning, problem-identifying and problem-solving, higher level of abstraction, system thinking, creativity, experimentation, collaboration, teamwork, leadership, effective communication, values of global citizenship, and ambition.

Curating Knowledge

The 'curriculum of the future' needs to curate knowledge by adopting interdisciplinary and transdisciplinary learning to enhance systems thinking that allows connecting knowledge from internal and external environments to better inform and understand complex problems.

Although it has many elaborations, system thinking is a dynamic process that leading systems theorists broadly defined as a way of thinking characterized by seeking a holistic view of issues [30]. This means that instead of looking at things as they are, we look at things in relation to the whole part that they form and not as individual parts or silos. In his book, 'The Fifth Discipline', Senge [31] provides an excellent introduction to the essentials and application of systems theory and how it can be brought together with other tools and devices to comprehend the whole and examine its interrelated parts. For Reynolds [32], another system thinking expert, "the traps of non-systems thinking can be observed in two dimensions: firstly, avoiding the inevitable interconnectivity between variables—the trap of reductionism, and secondly, working based on a single unquestioning perspective—the trap of dogmatism".

A reductionist way of thinking would ignore interconnections and assume a single cause [26]. Reductionism has been the leading paradigm through which we have understood the world to this day. This mode of understanding and thinking has led to assumptions that permeate all aspects of how we think about, manage, and design responses and solutions to problems. The focus on sustainability requires connectivity, synergies, and non-linearity, which entails a change in the way we, as a society, are collectively thinking about and understand the world around us to develop problem-driven contextualized responses that are aligned with the demands of the SDE. Systems thinking capability consists of an analytical skillset (the ability to see larger systems) and methodological tools (cause and effect—feedback loops, etc.), including modelling approaches.

While many soft and hard tools can apply systems thinking in practice, the essential ingredient in building the capacity to see the totality pieces of any system from a bird's eye or 'mega view'. Since we tend to operate in silo environments, seeing the whole picture is difficult as our systems do not necessarily provide all the information available to all players. For this specific reason, the 'curriculum of the future' needs to harness expert-guided yet self-directed multidisciplinary knowledge acquisition and problem-driven content that enables students to perceive the big picture and untangle the complexity of the challenges they are dealing with. Courses should be designed so that one discipline learns from the perspective of another and where the disciplines are integrated without their integrity being compromised. This allows learning experiences to be more context-specific, innovative, impactful, and of intellectual, practical, and vocational value to better prepare students for the SDE.

Next-generation leaders need to take fuller advantage of data and research science across several disciplines to uncover new insights, inform approaches, and move beyond past facts and assertions. This can better guide, enhance, and integrate the relevant dimensions of sustainable development knowledge.

Connecting Networks

The 'curriculum of the future' needs to draw on the knowledge and capacities of various individuals and types of collaborative platforms to build and link multi-stakeholder coalitions to co-create sustainable development actions. The capability to form powerful coalitions to drive systemic change is essential. One only has to turn to the preamble of the 2030 Agenda to illustrate the critical importance of such collaboration:

All countries and all stakeholders, acting in collaborative partnership, will implement this plan to free the human race from the tyranny of poverty and want and to heal and secure our planet. [33]

The related SDG17 seeks to strengthen the global partnership for sustainable development, complemented by multi-stakeholder partnerships that mobilize and share knowledge, expertise, technology, and financial resources. Forming such coalitions will rely heavily on individuals who share the commitment to deliver on the 2030 Agenda, yet in different ways. To gather a critical mass of individuals who will advance the same narrative and who can translate promises into action calls for creating an enabling environment for collaboration, which in turn requires a shift in the way, we inform and bring stakeholders together.

HEIs should play that role by increasingly involving academia, industry, government, and the community in co-creating learning experiences. For that purpose, universities need to embrace new practices through what the Organization for Economic Cooperation and Development (OECD) calls 'entrepreneurial universities [34]'. These universities consider entrepreneurship one of their strategic priorities and thrive on fostering novel relationships between internal stakeholders and with industry, society, and the public sector to create and transfer knowledge to solve humanity's global challenges. This goes beyond universities developing entrepreneurship and innovation centres, incubators, and science parks to facilitate the creation and growth of innovation-based companies through incubation and spin-off processes. Establishing effective entrepreneurial universities requires HEIs to develop strong connections with the world by committing to solving global challenges at the centre of the academic programmes.

In the upcoming sections, I present implementation case examples of the ACT model in an experiential global UN leadership programme and a university and further reflect upon these experiences to consider its utility for HEIs in contributing to sustainable development solutions.

3.3 Insights from the Application of the ACT Model in the UN Context

Development work is highly political. From a practical perspective, it is about who has access to resources and decision-making power. Navigating this terrain is messy, particularly when textured by human rights issues, social justice, democracy, exclusion, and conflicts. It is a constant balancing act between the normative and pragmatic, between changing people's hearts and minds, and cultivating productive relationships to influence in the right way to move the needle of progress forward. Achieving this balance is about choosing the correct sequence of leadership actions.

Throughout my years at the United Nations,[1] I have cumulated observations from my work with senior leadership of UN country operations including hot spots such as Yemen and Iraq. I have further led over two dozen global field missions to facilitate multistakeholder consultations between the UN and national partners on UN development priorities. As a UN instructor and trainer, I conducted and analyzed hundreds of learning needs assessments of leaders across the UN system as well as individual interviews with newly appointed country representatives of the Secretary-General as part of their induction programmes to prepare them for their leadership roles. The outcomes of these assessments not only informed the design of UN leadership development programmes but also guided leadership policy instruments. From the insights amassed, it was evident that the transition and the substantive paradigm shift from the MDGs to the SDGs required UN field leaders and their teams to possess different skillsets and capabilities to understand and address the dilemmas and changes brought on by the sustainable development agenda. This approach to delivering on the 2030 Agenda meant shedding outdated project-based thinking and siloed setups and exploring adaptive and collaborative solutions.

In the context of the UN country-level development operations, achieving the SDGs requires a UN Country Teams (UNCT)—comprised of senior representatives of all UN entities leading their respective operations in a country—to work beyond institutional mandates and move into practices of adaptive and collective leadership. This would allow them to promote co-creative directions and serve as enablers where the leadership and ownership of development results ultimately rest with the government and the citizenry. This transformation signalled that systems leadership capabilities are needed to do development work differently, demonstrate the UN's value proposition in the SDE, and contribute to the achievement of the SDGs.

The SDG framework requires understanding development priorities from multiple perspectives and considering effects among sectors and collaboration across disciplines to find solutions. The SDGs were different from the MDGs in three ways: they are universal and apply to every nation and every sector; they are all interconnected in a system, and they are transformative in recognizing that achieving them requires fundamental changes to how we live on Earth. Although laudable, the integrated nature of these attributes posed several compounding challenges, specifically for UN leaders tasked to drive sustainable development support at the country level. It required them to be knowledgeable in identifying and applying integrated solutions at the country level as mandated by the United Nations Secretary-General [35] and articulated in the United Nations Leadership Model

---

[1] Of the almost 20 years working in United Nations headquarters and field operations, dedicated over a decade to leading in-country multistakeholder consultations on development and humanitarian priorities amongst governments and UN country teams; strengthening senior and emerging UN leadership capacity at the UN System Staff College through the design, delivery, and evaluation of tailored leadership development programmes; leading and advocating at UN Headquarters for new system-wide leadership policies and instruments to support the strategic repositioning of UN country-level engagement; and creating and testing transformational leadership approaches and learning strategies to ensure UN field leaders are fit-for-purpose.

[36]. Taken as knowledge 'hardware', these are critical components to lead successful sustainable development efforts and create pathways to build healthy societies for generations to come. As ambitious as those sounds, the notion suggests if the trajectory of sustainable development requires addressing complex problems, then the way to address these problems should change fundamentally towards a system learning approach that builds the capabilities needed to navigate complexity.

The essence of systems leadership is the ability to identify those dynamics, manage relationships among different groups, and bring all key stakeholders to form a new commitment for whole-of-society change to drive the aspirations of the 2030 Agenda. To be successful, this SDG-driven multi-stakeholder process requires a coalition, a common problem-driven approach, a platform, a shared language, and a 'holding space' that allows stakeholders to work on complex and at times conflicting issues in constructive, experimental, and co-creative ways. Scharmer describes this as the formation of a 'collective container' wherein emerging impulses for the future can be heard 'in yourself, in others, and between you [37]'. It is broadly acknowledged that experimentation spaces need to be established to understand better what buttons to push for system change [38].

To deliver on the aspirations of the SDGs, in 2018, I designed and field-tested 'SDG Leadership Labs' to provide a 'holding space' for the UNCTs to explore adaptive challenges, collaborate around ideas, and prototype solutions together with national partners.

3.3.1 SDG Leadership Labs

As the United Nation's first experiential labs focused on leadership solutions for sustainable development, the SDG Leadership Labs were designed to promote systems thinking and cultivate a mindset of doing development work differently. Essential to the lab concept was tackling real, sustainable development issues in line with the regenerative leadership systems approach of the ACT model, with a focus on individual and collective leadership.

The lab itself was conceived to be adaptive and collaborative in its design. This built-in flexibility allowed for tailoring to the specific needs of UNCTs and engaging potential collaborating partners to apply their respective systems methodologies and tools in line with the ACT model components and action principles. The Uganda and Cambodia UN teams were selected to pioneer the lab approach. A partnership was struck with MIT senior lecturer Otto Scharmer and his Presencing Institute to apply his 'Theory U' systems methodology. This approach aimed to promote individual mindset and group behaviour change to enhance multistakeholder collaboration through new ways of thinking and working. The lab delivered a 4-month leadership journey for the UN teams through onsite experiences to explore adaptive and collaborative capabilities and apply co-creative tools to prototype solutions to real challenges they were contending with. The lab offered a solutions space to transform how the UN can strategically collaborate with new quadruple helix partners through systems sensing and transformation.

For the Uganda lab experience, collaboration was strengthened amongst members of a newly formed UNCT through a mutual understanding of systems approaches in the context of UN development planning and implementation and its drivers. The Cambodia lab established a systems architecture to reposition the UN system to implement new strategies for multistakeholder collaboration [39]. One SDG Leadership Lab participant explained the experience:

> The Lab helped us to get out of the old sectoral, siloed approaches we have been used to implementing for so long. I believe this leadership journey can certainly guide us to work closely together with new tools to do more adaptive development, which is not so natural, as we are used to doing things in a certain way. We should address the complex problems that our systems have created, instead of solving just symptoms, thus providing more integrated and effective solutions to the challenges of the SDGs. [39]

The SDG Leadership Labs provided a permissive environment to experiment with new ideas and systems methods without the risk of UN development operations deviating from course and purpose. Furthermore, it stimulated new thinking outside conventional practices and operational comfort zones to address complex development challenges and support leaders as they iterate, stumble, and adapt to find new solutions with national partners [24].

3.4 Extending the ACT Model to Higher Education

UNESCO's recent proposal for a new social contract for education, grounded in a commitment to education as a public societal endeavour and a global common good, is a call to rally. In the years leading up to this, there were plenty of signals of enthusiasm and interest reverberating through the corridors of HEIs that were pursuing dimensions of ESD and the exhaustive coverage of the topic across the internet. A simple Google search for 'ESD in higher education' yielded 631,000,000 results on February 12th, 2022. In his article, Karatzoglou [40] efficiently offers a quasi-experimental research analysis of the literature reviews of university experiences and contributions to ESD, which he reports are brimming with 'case-study type' articles on HEIs engaging with sustainability in descriptive and at times 'instructive, problem-solving terms'. His analysis marches on to suggest that the case studies predominantly illustrated success stories that described general patterns of how academics and stakeholders have collaborated yet did not find cases of systematic learning and engagement. Earlier in the chapter, we embraced ESD as a learning process based on the ideals and principles that underlie sustainability. And taking ESD as a transformative process of adaptive and collective praxis throughout changing social, economic, and environmental circumstances, the value of the ACT model in promoting regenerative learning becomes evident.

Senge [31] suggests, albeit taken at the organizational level, that 'adaptive learning' must be joined by 'generative learning'—learning that enhances the capacity to create. And Oakeshott [41] points out that education is not just desirable but integral to building sustainability. In practical terms, not one model or methodology is sufficient alone in tackling matters of complexity at all levels across any dilemma or challenge within any context. To develop multidimensional systems thinkers, HEIs need to curate knowledge and be more cohesive and integrated across their programme disciplines to expand 'whole-of-system' expertise to students to solve the increasingly complex problems of achieving sustainability. This would require revitalized internal coordination mechanisms and the ability to demonstrate the value and impact of interdisciplinary learning that ensures the knowledge higher education generates is anchored in social, economic, and environmental justice [16]. HEIs should also ensure that students can draw on a steady stream of networked multisectoral insights and intelligence from local, regional, and international actors. The ACT model would provide a 'knowledge to know-how continuum' for HEIs in preparing future leaders. It would help them to better (i) adapt thinking and behaviours that enhance experimentation and risk-taking in response to constantly changing environments; (ii) collaborate to seek a shared understanding of problems, enabling joined-up support for practical solutions to challenges; and (iii) transform themselves and their society's focus on positive change within their communities.

3.4.1 The United Arab Emirates University Pathfinders Programme

The most recent application of the ACT model is in the United Arab Emirates University's (UAEU) innovative and entrepreneurial programme on preparing the next generation of students to become *job creators* instead of *job seekers*. In 2021, the UAEU piloted its ground-breaking 'Pathfinders Programme' [42] as part of its 'Future of Education' project, launched at World Expo 2020 in Dubai [43]. The programme is the prototype initiative to develop an interdisciplinary *Bachelor in Impact Driven Innovation*. The programme is based on students being immersed in learning essential skills and capabilities through modules (pathfinding phase) and acquiring multi-dimensional knowledge and insights from subject matter experts to understand the challenges at hand (immersion phase) and then concept solutions to four global macro-challenges (conception phase). The four macro-challenge categories for a sustainable life were: (a) healthcare and wellbeing; (b) space exploration, biotechnology, and artificial intelligence; (c) food and water security (with a focus on the environment ecosystem); and (d) building a fair and just society (with focus on poverty alleviation, job creation, and socioeconomic growth).

The programme is an expert-guided, self-driven, inquiry-based learning journey for students. I led the UAEU's adaptation of the ACT model into its future-focused curriculum under 'leadership and collective action' to drive adaptation, collaboration, and transformational change needed to generate sustainable development solutions. The module provided students with a 'holding space' to apply the ACT model and introduce students to transformational learning through systems thinking. Specifically, students explored the context and challenge space in-depth and sought to reimagine possibilities and improve and redefine current resources, processes, constraints, boundaries, and assumptions. The learning experience included curated sector-specific and theoretical knowledge, systems thinking and adaptive skills development, engagement with a network of industry players and practitioners, and cultivating a personal and group perspective. The programme continued with participants exploring the solutions space in experiential conditions to develop and test prototyped solutions.

Through the regenerative leadership system approach of the ACT model, students experienced personal transformation as natural and modular self-expression that can continuously adapt to changes in the system. For the students, the module allowed for a multi-level understanding of being who they need to be as leaders and what it takes to exercise effective collaborative leadership when prototyping their solutions to macro-challenges. The ACT model proved to be a relevant theoretical framework that narrowed the divide from theory to praxis by cultivating capabilities for multilevel systems thinking. Students learned to seek opportunities for action in the real world and discover pathways for individual growth through creating possibilities for their ideas and innovations.

4 Concluding Comments and Reflections

The fourth goal of the 17 SDGs focuses on education. It can be considered a force multiplier acting as a driver and accelerant of critical sustainable development conditions: health and prosperity, productive livelihoods, economic and employment security, and the full development of human potential [44].

In his 1916 seminal piece on education in *The Atlantic* journal, British philosopher and mathematician Bertrand Russell wrote, 'Education should not aim at a dead awareness of static facts, but at an activity directed toward the world that our efforts are to create [45]'. This statement encapsulates the vision of ESD to which HEIs need to explore and respond meaningfully.

Higher education plays a critical role in educating and preparing a new generation of leaders. Participants at the 2014 UNESCO World Conference on ESD agreed on this notion that has since advanced the call for reorienting HEIs to integrate leadership development into their curricula [46]. The importance of this exercise cannot be stressed enough.

The first part of this chapter presented the prominence of ESD within the evolution of the SDE. The section briefly highlighted the progress towards sustainability and the SDGs, the leadership challenges in adapting to the demands of the 2030 Agenda, and the critical role and contribution of higher education in preparing leaders to be both learners and practitioners of the ESD process. The second part of the chapter introduced how a 'University of the Future' can contribute to the province of sustainable development by creating space to tackle complex problems through solution-focused approaches. To do this, the chapter offered specific measures on how HEIs can reorient focus and embrace capabilities development to elaborate a 'curriculum of the future' grounded in systems leadership.

The chapter introduced three action principles that would guide curriculum development and implementation cultivating capabilities, curating knowledge, and connecting networks—and promote adaptation and collaboration that underpin individual and collective transformations through praxis. Taken together as new leadership 'software', these principles and their components constitute a regenerative leadership system for sustainable development called the 'ACT Model' (Adapt, Collaborate and Transform). The chapter further illustrates the application of the ACT model in the United Nations field leadership development initiative called 'SDG Leadership Labs' launched in 2018 and the UAE University's Pathfinders Programme for its 2021 'Future of Education' project. These initiatives are discussed in the context of the potential utility of the ACT model to HEIs in graduating future-fit leaders for the SDE—and beyond.

The ACT model and the case examples in this chapter place leadership capabilities development within a living set of dynamics and applications. These influence a student's capacity and knowledge to expand in connection to others by highlighting how agency is shared as well as the diverse and networked dimensions of knowledge itself [17]. The ACT model's value emerges when the multiple aspects of the system work together in a regenerative manner, with each part playing its role and supporting the others to provide the primer for individual and collective transformation through adaptive and collaborative actions. The model recognizes that transformation is not a linear process but dependent on a series of systems that reinforce one another and consider praxis as an essential fulcrum for cultivating systems thinkers and users who can meaningfully engage in a complex world.

HEIs are best placed to contextualize and implement the ACT model according to their context, resources, and challenges. Among the initiatives that could be adopted to implement the ACT model, HEIs could establish a holding space such as an integrated leadership capabilities lab or academy structure. This mechanism would be accessible to all students across university colleges and departments. A challenge-based interdisciplinary and transdisciplinary programme related to the SDGs could be offered as a capstone course or micro-credentials. The curriculum design in this programme could be guided by the 3C action principles discussed in this chapter. Lastly, by involving the quadruple helix actors in the programme, HEIs could apply and test a regenerative leadership system inspired by the ACT model. It is important to note that the model's utility to address complex problems extends to any connected networks or actors of the quadruple helix—academia, government, industry, and community—and beyond.

Although HEIs are major places to 'learn' about lifelong learning, they should not be the only ones to play a critical role in preparing future leaders with sustainability knowledge and leadership capabilities. ESD should begin earlier at primary and secondary levels to better prime students to navigate life's changes, seek possibilities for growth and create abilities to connect new knowledge and ideas on sustainability. It is worth glancing back at target 4.7 of SDG4 mentioned earlier in the chapter, which maintains that *all* learners must acquire the knowledge and skills needed to promote sustainable development. The significance of ESD to the younger generation extends beyond the achievement of aspirational global education targets. Cultivating sustainability mindsets and behaviours of students at early educational levels can set out a foundation ground stone to build up basic knowledge and understanding of sustainable lifestyles, human rights, climate change, the natural environment, promotion of peace, appreciation of cultural diversity, gender equality, and global citizenship. In 2019, Italy was a first mover in this effort as the first country to begin incorporating the 2030 Agenda for Sustainable Development in the curriculum for schoolchildren and making sustainability and climate crisis a central part of its education model [47].

ESD is an ongoing and evolving process of learning and discovery that should be implemented widely by different players and at varying levels of education and society to drive action and build prosperity for people and the planet. The next generation of leaders is guardians of our natural world. Preparing them to sense and adapt to changes to better contend with complexities would help them foster a richer sense of who they are, what they value and where to create it for themselves and others, and how and with whom they want to move forward in the world. The moment to 'ACT' is now.

References

1. Stiglitz, J., Sen, A., & Fitoussi, J. P. (2009). *Report by the Commission on the measurement of economic performance and social progress* (pp. 11–14). OECD.
2. Adams, W. M. (2009). *Green development: Environment and sustainability in a developing world* (3rd ed.). Routledge.
3. World Commission on Environment and Development (WCED). (1987). *Our common future* (p. 41). Oxford University Press.
4. World Commission on Environment and Development (WCED). (1987). *Our common future* (p. 43). Oxford University Press.
5. United Nations Conference on Environment & Development. (1992, June 3–14). *Agenda 21*. http://www.un-documents.net/agenda21.htm. Last accessed 15 Feb 2022.
6. United Nations Economic and Social Council: Implementing Agenda 21. (2002). *Report of the Secretary-General. Commission on Sustainable Development acting as the preparatory committee for the World Summit on Sustainable Development, Second preparatory session*.
7. Chabrak, N., & Richard, J. (2015). The Corporate World & sustainability: Eco-efficiency & the doxic shareholder value. In G. Aras (Ed.), *Sustainable markets for sustainable business*. Gower. ISBN: 978-1-4724-3341-1. http://www.gowerpublishing.com/isbn/9781472433411
8. United Nations Environment Programme. (2016). *The financial system we need: Aligning the financial system with sustainable development*. ISBN: 9789210602426. https://www.un-ilibrary.org/content/books/9789210602426
9. Zilahy, G., & Huisingh, D. (2009). The roles of academia in regional sustainability initiatives. *Journal of Cleaner Production, 17*(12), 1053–1056.
10. United Nations Conference on Environment & Development. *Agenda 21, chapter 36: Promoting education, public awareness and training*. http://www.un-documents.net/a21-36.htm. Last accessed 15 Feb 2022.
11. UNESCO. *What is education for sustainable development?* https://en.unesco.org/themes/education-sustainable-development/what-is-esd. Last accessed 15 Feb 2022.
12. Wals, A. (2009). *Review of contexts and structures for education for sustainable development: Learning for a sustainable world. UNDESD 2005-2014*. UNESCO.
13. United Nations Sustainable Development Goals. *SDG 17: Revitalize the global partnership for sustainable development*. https://www.un.org/sustainabledevelopment/globalpartnerships/. Last accessed 15 Feb 2022.

14. UN SDG Learn Platform. *United Nations SDG primer*. https://www.unsdglearn.org/courses/sdg-primer-e-course/. Last accessed 15 Feb 2022.
15. G20: The G20 Education Ministers' Declaration: Building consensus for fair and sustainable development: Unleashing people's potential. (2018). http://www.g20.utoronto.ca/2018/2018-09-05-g20_education_ministers_declaration_english.pdf. Last accessed 15 Feb 2022.
16. World Economic Forum (WEF). (2022). *Expert's network brief on education, skills and learning*. Curated by Prof. Dr. Isabell M. Welpe and Felix Rank. Technical University of Munich.
17. United Nations. (2021). *Our common Agenda—Report of the Secretary-General*. https://www.un.org/en/content/common-agenda-report/. Last accessed 15 Feb 2022.
18. United Nations Educational, Scientific and Cultural Organization (UNESCO). (2021). *Reimagining our futures together: A new social contract for education*. International Commission on the Futures of Education. https://unesdoc.unesco.org/ark:/48223/pf0000379707. Last accessed 15 Feb 2022.
19. Leading in Uncertainty: Using the Cynefin Framework to Excel as Leader. *Canada School of Public Service, Government of Canada*. https://www.csps-efpc.gc.ca/tools/jobaids/lead-cynefin-eng.aspx
20. European Commission, Joint Research Centre, Rancati, A., & Snowden, D. (2021). *Managing complexity (and chaos) in times of crisis: A field guide for decision makers inspired by the Cynefin framework*. Publications Office. https://data.europa.eu/doi/10.2760/164392
21. Churchill, W. (1942, May 5). *Memo: Piers for use on beaches*. https://www.iwm.org.uk/collections/item/object/205195425. Last accessed 15 Feb 2022.
22. Mulberry Harbour and PLUTO. *Narrated by John Newsome, Institution of Civilian Engineers*. https://www.youtube.com/watch?v=DXF1ov0KAQE. Last accessed 15 Feb 2022.
23. Trueman, C. N. (2015, April 21). *The Mulberry Harbour. The history learning site*. www.historylearningsite.co.uk. Last accessed 15 Feb 2022.
24. Khan, S., & Papoulidis, J. (2019, April 4). *A new role for UN leadership in the hardest places*. Devex Global Views. https://www.devex.com/news/opinion-a-new-role-for-un-leadership-in-the-hardest-places-94554. Last accessed 15 Feb 2022.
25. Giddens, A. (1976). *New rules of sociological method*. Hutchinson.
26. Bourdieu, P. (1977). *Outline of a theory of practice*. Cambridge University Press.
27. Arendt, H. *The human condition* (2nd ed.). Chicago University Press.
28. Gramsci, A. (1999). Selections from the prison notebooks. In Q. Hoare & S. Nowell (Eds.) *Selections from the prison notebooks of Antonio Gramsci*. ElecBook. https://abahlali.org/files/gramsci.pdf
29. World Economic Forum. (2016, January). *The future of jobs: Employment, skills and workforce strategy for the Fourth Industrial Revolution. Report*. https://www3.weforum.org/docs/WEF_Future_of_Jobs.pdf. Last accessed 15 Feb 2022.
30. Systems Academy. *Systems*. https://systemsacademy.io/systems/. Last accessed 15 Feb 2022.
31. Senge, P. M. (1990). *The fifth discipline: The art and practice of the learning organization*. Random House.
32. Reynolds, M., & Howell, S. (2010). *Systems approaches to managing change: A practical guide* (pp. 1–23). Springer/Sci-Tech/Trade.
33. United Nations General Assembly Resolution A/RES/70/1. (2015). *Transforming our world: The 2030 agenda for sustainable development*. Resolution adopted by the General Assembly on 25 September 2015.
34. Organization for Economic Co-operation and Development (OECD). *A guiding framework for entrepreneurial universities*. Final version 18 December 2012. Last accessed 15 Feb 2022.
35. United Nations: Report of the Secretary-General. (2017, June 17). *Repositioning the UN development system to deliver on the 2030 Agenda—Ensuring a better future for all*.
36. United Nations Development Group. *United Nations Leadership Model*. Endorsed by the UNDG on 11 October 2014. https://unsdg.un.org/sites/default/files/UN-Leadership-Model-Rev-Jun-2017.pdf. Last accessed 15 Feb 2022.

37. Scharmer, C. O. (2009). *Theory U. Leading from the future as it emerges. The social technology of presencing.* Berrett-Koehler/McGraw-Hill.
38. United Nations Educational, Scientific and Cultural Organization (UNESCO). (2014). *World conference on ESD—Reports on workshops in Cluster IV: Setting the Agenda for ESD beyond.*
39. Buell, B., & Srivastava, M. (2019). *SDG leadership labs: Leading transformative change in the United Nations.* Presencing Institute. https://medium.com/presencing-institute-blog/sdg-leadership-labs-leading-transformative-change-in-the-united-nations-e41cbd35bc82. Last accessed 15 Feb 2022.
40. Karatzoglou, B. (2013). An in-depth literature review of the evolving roles and contributions of universities to Education for Sustainable Development. *Journal of Cleaner Production, 49*, 44–53.
41. Oakshott, M. (1989). *The voice of liberal learning.* T. Fuller (Ed.). Yale University press. Reprinted 2001. Liberty Fund.
42. UAE University. *University of the future retreat: Co-creating and co-designing the Pathfinder's journey.* https://expo2020.uaeu.ac.ae/en/retreat.shtml. Last accessed 15 Feb 2022.
43. UAE University. *Discover the future of Education.* https://expo2020.uaeu.ac.ae/en/index.shtml. Last accessed 15 Feb 2022.
44. Narayan, P. S. Achieving our education goals can unlock all the SDGs. WEF. https://www.weforum.org/agenda/2017/09/achieving-our-education-goals-can-unlock-all-the-sdgs/. Last accessed 15 Feb 2022.
45. Russell, B. (1916, June). Education as a political institution. *The Atlantic Monthly*, pp. 750–757. https://www.theatlantic.com/magazine/archive/1916/06/education-as-a-political-institution/305258/.
46. United Nations Educational, Scientific and Cultural Organization (UNESCO). (2014). *World conference on education for sustainable development conference.* Report by the General Rapporteur Heila Lotz-Sisitka, Professor. Rhodes University.
47. Hodel, K. (2019). Italy to put sustainability and climate at the heart of learning in schools. Global Education. *The Guardian.* https://www.theguardian.com/global-development/2019/nov/06/italy-to-school-students-in-sustainability-and-climate-crisis

Changes Required in Education to Prepare Students for the Future

Lobna A. Okashah, Akram Hamid, Jiwon Kim, and Ethan Rubin

1 Introduction

This chapter is broken into five sections. The first section will present the requirements for early education. The second section will show the history of changes in higher education, specifically engineering, to meet the demands of the society. The third section cites a civil engineering professor at Princeton University integrating different topics for an interdisciplinary approach to prepare her graduating students to work in city planning. The fourth section will present what the author sees as new trends needed in higher education based on Samsung's reports on 6G, Industry revolution 4.0, and the author's own professional experience. Finally, the fifth section will introduce how to sustain higher education to meet the demand of the future.

2 New Requirements for Early Education

The next phase of the industrial revolution will introduce artificial intelligence (AI) into all aspects of daily and professional life: machinery, transportation, healthcare, and entertainment will all utilize the Web 3.0 technologies that Generation Z has been using since childhood. If Qatar's education system embraces these developments, the next generation of high school and college graduates will be better

---

L. A. Okashah (✉) · A. Hamid
Kuwait Foundation for the Advancement of Sciences, Kuwait City, Kuwait
e-mail: akram@dallata.ca

J. Kim
Kuwait City, Kuwait

E. Rubin
Cypress, TX, USA

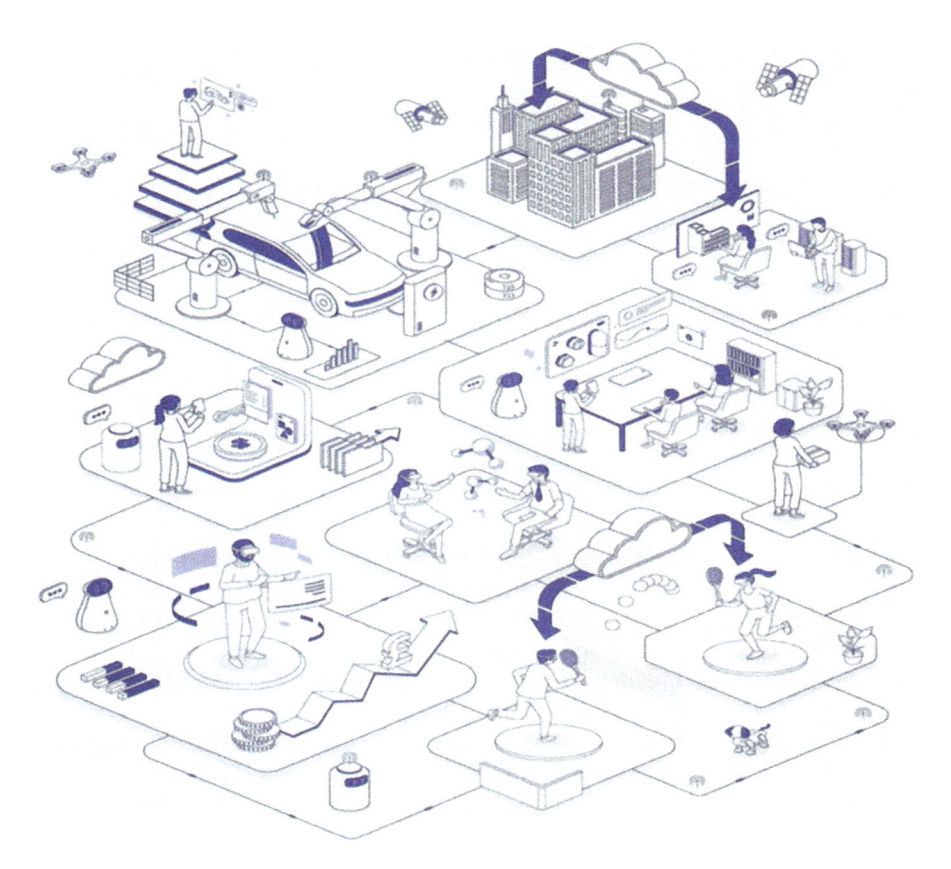

Fig. 1 Samsung 6G city

equipped to enter the workforce and create the software and infrastructure needed to realize Qatar's planned smart cities. Samsung's proposed 6G cities use various novel technologies, which in the future will be part of daily activities that will require 6G to operate smoothly (Fig. 1). It shows technologies such as Cloud computing, AI, robotics, VR, AR, data analytics, Bitcoin, non-terrestrial objects, etc. Such technologies must be integrated into the standard education at an early stage.

By combining extant curriculum and technological tools, even primary school students can learn more efficiently and effectively. The main themes of this chapter are intersectionality and technological integration, both of which involve meeting each child in their "comfort zone" to help them grow.

A personalized approach for each student is ambitious but is more possible today than ever before thanks to developments in education technology. Educational software can identify students' individual issues and tailor content to suit each student's needs [11]; deliberately targeting students' misconceptions, rather than giving feedback on a question-by-question basis has returned better results in test groups [2, 8].

Beyond the scope of the individual student, the pandemic has hindered students' social development and, as a result, their academic performance. In the 2020–2021 school year, students in the United States were on average 4 months behind in reading and 5 months behind in mathematics (Emma Dorn [3]). The report also showed increases in mental health conditions and behavioral issues such as anxiety and depression due to the pandemic. Despite concerns about students' time spent online, technology *can* help reverse these trends. Even before the pandemic, modern primary and secondary school students' social lives included a great deal of technology such as multiplayer video games and remote communication outside of school hours. Rather than drawing a firm line between their analog school life and their digital personal life, we can "meet them in the middle" and foster greater interest in their school subjects. Consider Qatari primary school students using virtual reality (VR) headsets to explore medieval sites in Europe after learning about the Arthurian legend: students get to use a technology they are familiar with, heighten their understanding of the literature, and the price of a VR headset and leasing scans of a site—many of which are available for free—is far cheaper than a ticket to England. Now, consider secondary students revisiting European texts in greater detail as teenagers after taking a programming class. For a group project, students could combine both subjects to write and program a text-based roleplaying game based on one of the stories they have read. In both examples, students are introduced to traditional educational materials and then enhance their understanding by using technology they may already use in daily life.

Implementing these changes will require involvement from and coordination between many entities that were previously unengaged. A comprehensive education model requires alignment of the goals and plans of both the government and the private sector. With this in mind, the suggested action plan covers the execution of *two parallel approaches* designed to educate students about Web 3.0 and 4.0 technology:

The First Approach Extracurricular activities focused on technology should be developed and made fully accessible. Children should be introduced to critical thinking, the basics of machine learning, AI, automation, game theory, programming, and multiple international languages. Curriculum must be comprehensive, age-appropriate, and keep up with real-time technological advancements and should utilize the same technologies used in the professional world and private sectors, such as VR and augmented reality (AR). This approach requires input and resources from the private sector and may include providing or leasing out proprietary technology for older students on the verge of entering the workforce.

The Second Approach Magnet schools should be opened with a wide variety of focuses, and they should be available to students based on interest rather than zoning or test scores to avoid discrimination [1]. When students transition from primary to secondary school, they should be given the opportunity to pursue the interests and skills they have already developed; students that excel at the arts should attend a school that fosters, rather than suppresses, their passion, and the same applies for more technical-oriented students. These schools should not be mono-subject but should instead teach a well-rounded curriculum centered around their specialty. Art-focused magnet schools, for example, should give design and sculpture students opportunities to study under and collaborate with civil engineers and architects; robotics-focused schools should include significant social studies and history education to prevent the lapses in education that have created discriminatory AI.

Qatar must seriously think about abolishing the current "one kind fits all" approach to education. With the slowdowns created by the pandemic, now is a good time for a large-scale restructuring. If Qatar can become a leader in individual-focused learning, it can be a leader for other countries to follow and draw in parents seeking better education for their children.

As mentioned in the first approach, schools are not the only places where children will learn. For example, museums, public libraries, and NGOs are all currently underutilized in public education. Curriculum for musically inclined students should be developed with input from Qatar's music and cultural organizations, so students can receive an education that meets what the industry wants. By embracing students' individual passions, we can also make it easier for them to learn outside of their comfort zone; for a musically inclined student, framing their higher-level physics instruction around sound can help them grapple with a difficult subject and improve their musical education.

Changes in education are crucial if Qatar is to progress on its plans for smart cities. Today's students will be the ones designing, building, and living in these cities; they must be prepared to do so. The following sections will explore some of the proposed elements of smart cities and the further changes necessary to prepare for them, along with how we can adapt the education system to meet those needs.

3 How Historical Precedent Informs the Future of Education

Figure 2 diagrams the changes in the field of engineering over the past two centuries. The needs of the private sector (e.g., industrial, farming, health, etc.) have long-driven scientific research, legislation, specification, and even market demands and consumers' needs. This development and specification are not unique to engineering; every field has undergone similar changes, but the engineering tree illustrates just how quickly the field diversified.

When discussing the logistics and infrastructure of proposed smart cities, Dr. Ramswami of Princeton University explained that her team studied cities and their sustainability by considering "shelter, water, food, energy, connectivity, sanitation, and green spaces" [14]. All of these elements involve an intersection of hard science and social science, and when describing the skills needed to do such analysis, Dr. Ramswani described "a new type of professional, who can combine concepts from urban planning and infrastructure engineering and look at both through a lens of industrial ecology or urban metabolism. That's the kind of students I've been

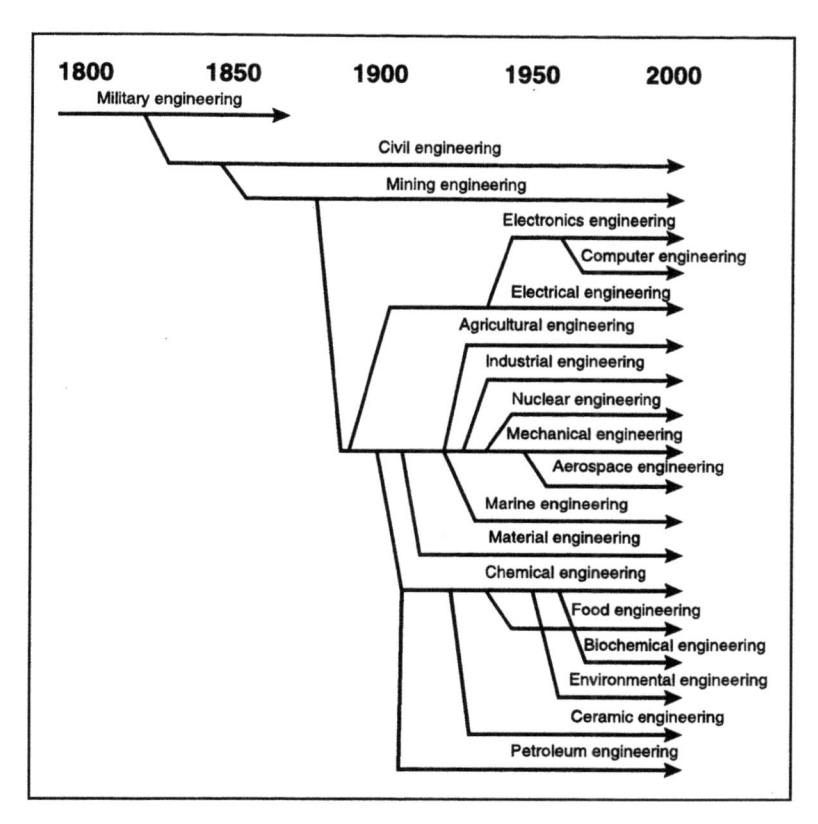

Changes Required in Education to Prepare Students for the Future 111

Fig. 2 A schematic approximation of the historical evolution of engineering disciplines. (Reprinted with permission from Tadmor [5])

training in some of my National Science Foundation-supported programs in the U.S." [14]. Based on these quotes, we know that developing and educating this type of intersectional student is possible but requires funding and intent. For tomorrow's smart cities, the traditional civil engineer must learn more than just the strength of material and the physics of concrete; they must learn about the complex world of social and logistical issues that influence a city. Eventually, this kind of education will become the norm in the developed world.

4 New Trends for Higher Education

The world has experienced four industrial revolutions so far; first in the eighteenth century with the invention of steam powered machinery; then in the nineteenth century with mass production and electricity; the third in the mid-twentieth century with computers, electronics, and information technology; and now, the fourth,

The Technology (R)evolution arrives in three waves of disruption

[Diagram: Industry 4.0 circular diagram with Physical World and Virtual World forming the Digital World, surrounded by 17 pillars: Robotics (M2H, H2M, M2M), Bioinformatics, Nanotechnology, 6G Communication, Blockchain, Internet of Things, Autonomous Systems, Cybersecurity, Neurotechnology, Smart Automation, Quantum Computing, 3D Printing, Future of Energy, Advanced Material, Augmented Reality, Artificial Intelligence, Advanced Analytics. © 2020 Henrik von Scheel™]

1st wave 2009 – 2016
- Advanced Analytics
- Augmented Reality
- Cloud Computing
- Digitalization (IOT)
- Robotics
- 3D Printing

2nd wave 2016 – 2025
- Artificial Intelligence
- Autonomous Systems
- Blockchain
- Smart Automation
- 6G Communications
- Web 3.0

3rd wave 2025 …
- Advanced Material
- Bioinformatics
- Cybersecurity
- Future of Energy
- Nanotechnology
- Neurotechnology
- Quantum Technology
- ……

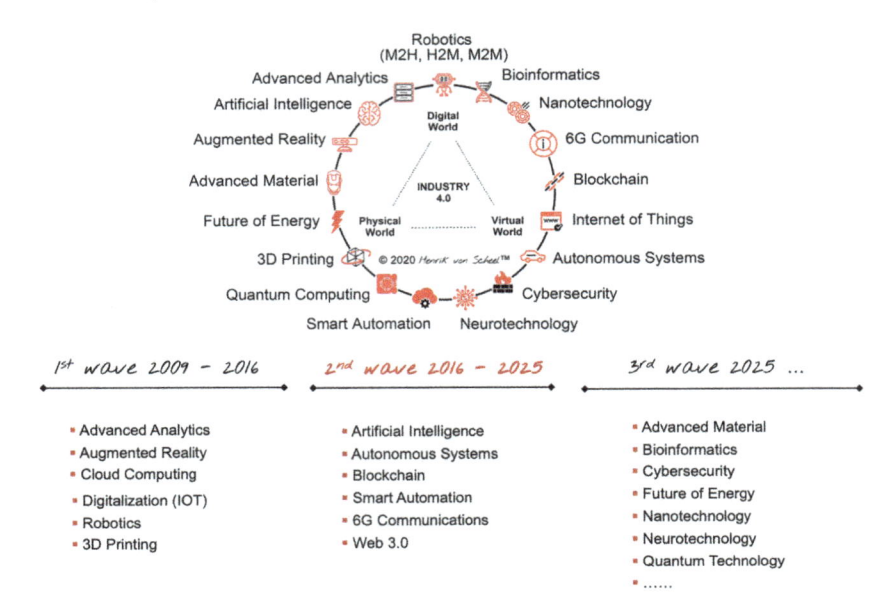

Fig. 3 von Scheel

through smart technology, described as a "fusion of technologies that is blurring the lines between physical, digital, and biological spheres" [13]. In Putting Industry 4.0 into Practice, Henrik and Joshua van Scheel outlined 17 distinct pillars of technology involved in blurring those lines (Fig. 3):

The authors posit that the fourth industrial revolution is arriving in waves, each of which focuses on a different "pillar." The first wave, staring in 2009 and continuing until 2016, emphasized 4 of the 17; the current wave focuses on AI, blockchain, automation, 6G, and energy. Using these pillars as a basis, this chapter will outline how to address some of these topics in higher education, especially in the medical and technological fields, though the potential scope of technological integration can and should include virtually every educational field.

The rapid changes that the globe is going through will require a more agile education system that can adapt to the changes faster than in the past. An article from the October 4th, 2021 issue of Inside Higher Education states that the future education will have to be "outcome based, time independent, digital, individualized, low cost, available any time and any place" [6]. Those reforms can be achieved using new technology such as extended reality (XR), which is an immersive technology, and VR and AR. AI and robotics are the major technology that will have to be utilized in education, as well as Holograms that will be utilized for distant learning by having a human-sized hologram of the instructor present in front of the students.

VR, AR, and XR have already been used to fast-track training in the energy sector. In 2021, the International Trade Council (ITC) reported that a program used for training deep sea divers to install underwater pipelines at a depth of 2000 m, which previously took a year and a half to complete, was streamlined using XR instruction and allowed divers to complete their training in only 4 months, saving the company 3.2 billion dollars and making the training process far safer [4].

In the United States, Michigan Institute of Technology has developed the XR programs *Cellverse* and *Electrostatic Playground* to teach molecular biology and atomic physics, respectively, and Case Western Reserve University uses the software *HoloAnatomy* to teach anatomy. XR technology can be utilized in many disciplines, but its impact in medical science is especially notable. Using XR, students can simulate various surgeries with all sorts of complications, all without touching a single cadaver or patient. XR can make better doctors for cheaper and with less risk during the education process.

Medical students also need instruction to operate in a hybrid healthcare environment. Hybrid healthcare refers to the combination of traditional healthcare and remote digital ("telehealth") healthcare. Covid-19 has led to an explosion in telehealth to prevent transmission between patients and doctors, make healthcare more accessible in remote areas, and cut down on operating costs. The future of remote healthcare will involve AI-operated machines for sample collection and processing, so healthcare workers will need to be able to interface with their technology.

In the same vein, healthcare workers also need to have sufficient technological literacy to keep their patients safe. In an age rampant with digital misinformation, hospitals are increasingly vulnerable to ransomware and phishing attacks, and, unfortunately, staff are falling for them [9]. Technological literacy is not just an issue of improving care, but of preventing catastrophic system failures and information leaks.

Another pillar involving the medical field is 3D printing, which has transitioned from a novelty to a feasible means of prosthetic production. Here, sculpture, technology, and medicine meet to create something with real-world, life-changing potential, but it requires cooperation from all three fields; the sculptor may understand the software, but without the medical knowledge, they cannot create a valuable prosthetic or artificial valve, and vice versa.

With the increase in the number of applications that rely on AI, there will be more demand in the fields of algorithm programming, spatial intelligence, autonomous object, spatial and temporal reasoning to name a few of the fields that will need more research and therefore be taught in higher education institutes. The following graph, which was created by Joshua and Henrik van Scheel from their book "Putting Industry 4.0 into Practice," shows the trends in AI. It can be considered as a guideline to implement the areas that are lacking in the current curriculum and incorporate the more advanced ones in the future (Fig. 4).

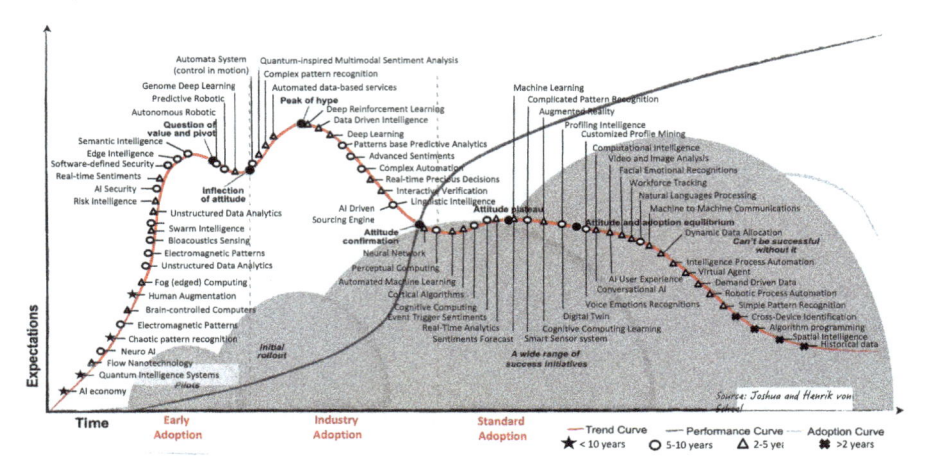

Fig. 4 AI disruptive trends

5 Requirements of the Secondary School Curriculum

With the potential benefits of technology in mind, we must develop curriculum that prepares students to create and utilize these assets, which will require significant infrastructure changes. The coming years will demand faster networks and wireless communication. Samsung has released a report about the technology needed to support a transition to 6G, which include terahertz (THz) bands, novel antenna technologies, evolution of duplex technology and network topology, spectrum sharing, comprehensive AI, split computing, and high-precision networks. Each of these requirements come with new demands, new training, and new jobs in sectors spanning from finance to transportation.

The THz bands needed for 6G require solid state electronics, new material for antennae, and new protocols to implement it all. 6G will also use non-terrestrial network (NTN) components such as satellites and high-altitude platforms for coverage in remote areas (see Fig. 5). These infrastructure elements will operate autonomously by using AI with special and temporal reasoning. Every aspect of this technology must be understood, developed, and maintained, all of which can be done by today's Qatari students if they are given comprehensive education.

All proposed smart cities must contend with the geographical threats of their region, which are becoming more severe due to climate change. Competent civil engineering has, for example, created "Sponge City" in China, where an area that floods frequently has been designed with an innovative drainage system, absorbent

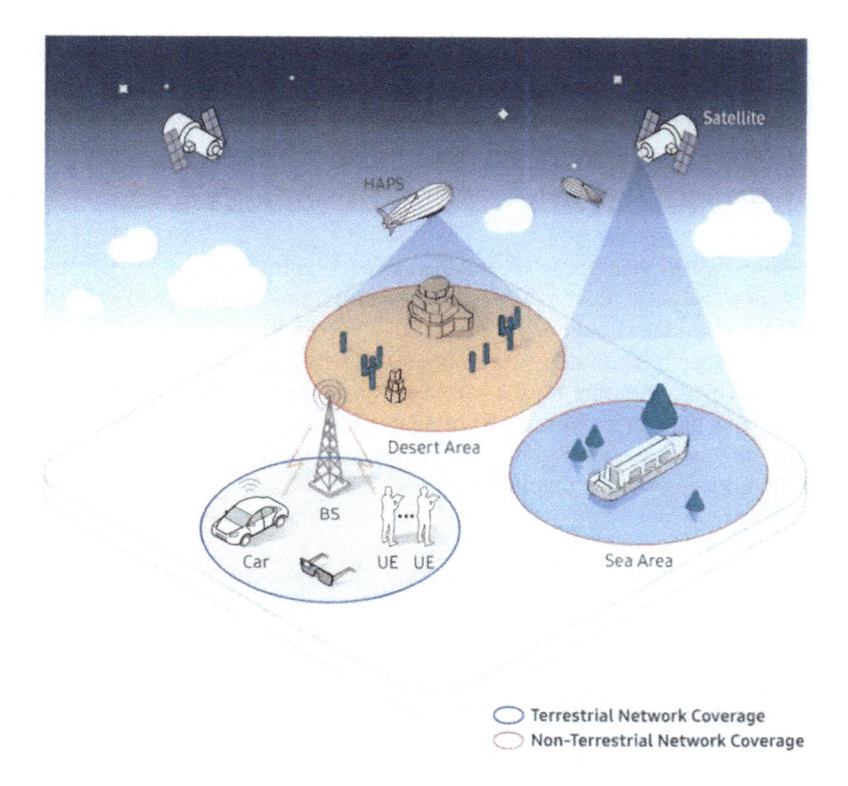

Fig. 5 6G non-terrestrial network

material, a robust water recycling system, and integrated water storage in rain gardens and artificial water features. This is the kind of intersectional city planning possible only when students are educated in ecology, materials science, geology, and design [7].

With all this comes a tremendous amount of data, which students should be trained to analyze. All disciplines and sectors benefit from data analytics. We are in an era where disruption and innovation have created companies that have not even owned the commodity that they offer. Locally, drones are now being used to remotely monitor oil fields and pipelines without engineers taking a single step outside. Netflix is the largest film distributor but does not own a single movie theater. Alibaba is the world's largest retailer but does not have inventory. Talabat leads food delivery without owning a single restaurant.

None of these commercial successes would have been possible without soft skills, which many STEAM students lack when they graduate secondary school or university. Many graduates will have innovative ideas but no idea what to do with them; we can prevent this by teaching students about business models, how to write business plans, and the basics of intellectual property and registering patents.

Students should understand the various types of funding so they can make educated decisions about where to take their expertise.

6 Maintaining the Pace of Education

Every education system faces the challenge of keeping up to date. Technology changes constantly and drastically, so the curriculum must be designed with updates in mind. A committee should be formed with members from the private sector, public sector, research institutes, universities, and primary and secondary school authorities. This committee must meet every 6 months to discuss the technology trends and how the curriculum should be adapted. For example, with 6G technology in mind, students no longer need to learn about the now-obsolete technology unique to 2G. That does not mean that the curriculum is going to change every 6 months, rather to keep vigilant about the new surfacing technologies and create a reasonable and attainable plan to implement them.

The committee must be taken seriously by all parties concerned and should not be conducted last-minute or haphazardly. Delays in responding to new technology should be minimized; the private sector should consider this as an investment in their future employees. This committee will keep educational programs relevant and sustainable, and ideally motivate the private sector to invest and contribute.

7 Conclusion

This chapter outlined potential uses for technology in early and mid-level education, and how technology-focused instruction can replace or be given alongside the existing educational framework. By focusing on what society needs and wants as we transition to smart cities, we can adjust the educational system to produce graduates with education in these subjects. The chapter concludes by offering solutions for maintaining relevancy and sustainability in the proposed educational projects.

References

1. Bogen, M. (2019, May 6). Hiring algorithms can introduce bias. *Harvard Business Review*.
2. diSessa, A. A. (2006). A history of conceptual change research. In R. K. Sawyer (Ed.), *The Cambridge handbook of the learning sciences* (pp. 265–281). Cambridge University Press.
3. Dorn, E., Hancock, B., Sarakatsannis, J., & Virules, E. (2021, July). *Covid-19 and education: The lingering effects of unfinished learning*. McKinsey & Company.
4. International Trade Council. (2021, January 28). *Shaping the future*.
5. Tadmor, A. (2006). Redefining engineering disciplines for the twenty-first century. *The Bridge, 36*(2), 33–37.

6. Levine, A. (2021, October 4). The future of higher Ed is occurring at the margins. *Higher Education News*.
7. Li, H., Ding, L., Ren, M., Li, Ch., & Wang, H. (2017). Sponge city construction in China: A survey of the challenges and opportunities. *Water*.
8. Linn, M. C. (2006). The knowledge integration perspective on learning and instruction. In R. K. Sawyer (Ed.), *The Cambridge handbook of the learning sciences* (pp. 243–264). Cambridge University Press.
9. Rizzoni, F., Magalini, S., Casaroli, A., Mari, P., Dixon, M., & Coventry, L. (2022, January). Phishing simulation exercise in a large hospital: A case study. *Digital Health*.
10. Samsung Research. (2020). *6G the next hyper—Connected experience for all*.
11. Sawyer, R. K. (2008, May 15–16). Optimizing learning: Implications of learning sciences research. In *OECD/CERI international conference "Learning in the 21st Century: Research, Innovation and Policy*.
12. Scheel, v. H. (2019). *Putting industry 4.0 into practice*. Henrik and Joshua von Scheel Publishing.
13. Schwab, K. (2016, January 14). The Fourth Industrial Revolution: What it means, how to respond. *World Economic Forum*.
14. Spinney, L. (2021, June 12). Anu Ramswami Interview: How to shape the cities of the future. *New Scientist Magazine*.

Humanising Higher Education: University of the Future

Dzulkifli Abdul Razak and Abdul Rashid Moten

1 Introduction

In the Industry 4.0 environment coupled with the COVID-19 pandemic, there is a dire need for society to rescue higher education from its total reliance on Westernised (as per post-industrial revolution), economy-centred, industry-led, reputation-obsessed, and hence a dehumanising system. It should be replaced with a wholesome, inclusive, sustainable, equitable, and resilient (WISER) framework that will lead to a just and humane society with equal opportunities for all [1]. Otherwise, as predicted by some, humans will be overwhelmed by robotics and artificial intelligence (AI). The universities of the future must reclaim values and the human outcomes and realise human technology convergence by humanising technology. Based upon the content analysis of relevant materials, this study suggests the way by first humanising (higher) education in the context of the Fourth Industrial Revolution (IR 4.0) and beyond. It analyses the roles and functions of higher educational institutions and their contributions to the past, present, and future world of work and society.

D. Abdul Razak
Professor Emeritus Tan Sri Dato' Dzulkifli Abdul Razak is the Rector of the International Islamic University, Selangor, Malaysia

A. R. Moten (✉)
Professor Dr. Abdul Rashid Moten is the Guest Writer at the International Islamic University Malaysia, Kuala Lumpur, Malaysia

2 The Fourth Industrial Revolution

Governments in many parts of the world are investing huge resources into establishing higher educational institutions and hence there is a dramatic increase in enrolments around the world. By 2018, global post-secondary enrolment topped 200 million – up from 102 million in 2000 and this upward trend will continue throughout this century to promote technological and economic growth [2]. There has been an increase in the number of these institutions admitting more students than ever before. Yet, there persists an increasing scepticism about these institutions' contributions to the welfare of the society they supposedly serve [3]. Consequently, higher educational institutions are experiencing a major identity crisis requiring a reassessment of their roles and functions they are supposed to perform.

A major contributor to the identity crisis experienced by universities is the prevalence of neoliberal ideology that dictates and changes the fabric of our society. It prioritises "markets" and places "profit over people." It has monetised values and introduced markets and matrices into all spheres of life [4]. It has seriously affected the academic culture and landscape of higher education [5]. Knowledge therefore became a commodity, education became a business, and students emerged as customers, while graduates as products to be marketed. There is the drastic commercialisation and marketisation of universities. Rather than "conserving, understanding, extending and handing on to subsequent generations the intellectual, scientific, and artistic heritage of mankind" [3], universities have become a "factory" that churns out "human capital" for the economy and society [6]. Graduates should not be conceived as "human capital" in that such conceptualisation de-humanises education resulting from changes brought about by four industrial revolutions.

The First Industrial Revolution, with the invention of the Steam Engine in the 1780s, changed the workforce as the demand for manual labour dampened since machines were able to complete jobs faster and better. The Second Industrial Revolution, generally based in the period from 1860 to 1900, is associated with new manufacturing technologies leveraging on electricity, which triggered additional changes forging a "new economy." The Third Industrial Revolution is generally attributed to computerisation and web-based interconnectivity developed in the 1980s and the 1990s. This period witnessed an increase in access to higher education with greatly increased diversity on campuses and globalisation of academic research accelerated by online technologies.

The Fourth Industrial Revolution refers to the "fusion of technologies that is blurring the lines between the physical, digital and biological spheres" [7]. This fourth phase of the industrial revolution is characterised, among others, by exponential technological breakthroughs in the fields of robotics, artificial intelligence (AI), and the Internet of Things (IoT). It can be summed up as an inter-operable manufacturing process, integrated, adapted, optimised, and service-oriented.

One of the most visible effects of IR 4.0 is "Education 4.0," but it comes with its own set of problems [8]. The aim is to produce successful graduates for a world where these cyber-physical systems are prevalent across all industries. It gave rise, in a sense, to a "factory-like model" to mass produce "workers" for the industry. Education became a factory:

1. To create employment in the context of the workforce with an emphasis on employment and probability and marketability and so on. This requires mechanical-technical means,
2. To emphasise innovation, particularly technical innovation to train the mind.
3. To train knowledge as far as thinking is concerned, in order to create new things, get a better world, and also get a better economy for innovation.

These three objectives focus on manpower, mind, and machine, all geared mainly towards the Market. The three variables are interrelated from a macroeconomic perspective. Manpower enhances individual employment opportunities. It also serves as an input to produce innovation and therefore can impact technological change. Technological change can impact the development of labour demand both in quantitative and in qualitative terms. The employment consequences, in turn, depend on the available quantity and quality of human capital. Evidently, the sole concern of education is the economy and the development of an "enterprise culture" [9].

Students are offered practical knowledge called internship training. The aim is to back up the theoretical knowledge with the practical experience of the job. The university enters into an arrangement with industrial enterprises to provide hands-on knowledge to students who are internees. Industrial businesses get cheap labour for their extra work. The internship programme offers employment opportunities for students. Students, after graduation, are often employed by the organisation where they served as internees.

Academics and administrators working in universities are given KPIs (Key Performance Indicators) to evaluate their success and to be promoted to the next level. The indicators for academic staff, among others, include the research grants procured, a proxy measure of research relevance and competitiveness, and the number of refereed publications in reputable journals measuring the research output of a university. These requirements may force the staff to adopt immoral activities for upward movement. In their zeal for promotion, these KPI-driven academics often forget their obligations to serve society. Most of them publish in "predatory journals" by often paying prohibiting fees. This serves the author as well as the publisher. The author gets a chance for promotion and the "wealthy" publisher profits from it handsomely. Such immoral activities let the general public look down upon those working in the universities.

In the face of these criticisms, the proponents of the economic approach to education have changed their terminologies. Manpower is morphed into human capital, the mind becomes an invention, and the machine is turned into high technology. Not much has changed. Money and the market still remain the main concern of the education system. People compete with each other, sometimes they subdue their own fellow beings to be ahead of everybody else. It is all about one's own self, community, or country rather than humanity. This kind of education is called education without a soul. "The factory mass teaching methods of the third revolution era have failed to conquer enduring problems of inequity and unfairness" [10].

2.1 Education for Money

The human capital-invention-technology nexus is still very much the driver of education wherein there is no place for ecological and human dimensions. Ultimately, education is all about the logic of economics. It is about the value in monetary terms. A good job and a good salary are the defining characteristics of success in life. Universities and higher institutions must respond to the wishes of higher education clients. No attempts are made to change students' preferences but instead treat students as customers. Universities should offer the kinds of programmes students as customers want in a way that would maximise profit. The customer, after all, is always right. Education is to help students achieve their own personal pursuit of happiness, mostly in materialistic terms. The orientation of education in the contemporary world can be summarised in a five-letter word, WEIRD. It is Western-based, Economic-centric, Industry-led, Reputation-obsessed, and hence a Dehumanising exercise [11]. These are all deemed as the toxic consequences of the current higher education model [12] and are further borne out by the pandemic.

Educational institutions are under tremendous pressure from international businesses looking for workers with the requisite skills. They often conceived of universities as places to produce "human capital" for the market. Attempts are being made to restructure education that addresses the issues of modernity and to respond to the increasing demands of developing human capital rather than attending to those elements that give ordinary life depth, relatedness, and value.

This industry-led tendency has pressured universities in terms of their role and position. In *Excellence Without a Soul: How a Great University Forgot Education*, Harry Lewis (2006) laments that higher education has lost its sense of identity. Educational institutions allow competition and consumerism to drive their direction. Universities, designed to educate responsible human beings, have now evolved to a virtual "cafeteria model." They offer a diverse menu of disjointed options devoid of any intellectual integrity. "This superimposition of economic motivations on ivory-tower themes has exposed a university without a larger sense of educational purpose or a connection to its principal constituents" [13]. Instead of helping "students understand what it means to be human" [13], the universities now function within parameters of advancement, security, and reputation.

3 Education with a Soul

The need is for education with a soul. The soul has to do with authenticity/genuineness and depth of meaning, the loss of which would lead to troubles of all sorts individually and socially [14]. In this technological age with the prevalence of neoliberal ideology, the mind is separated from the body and spirituality is at odds with materialism. The soul is sorely needed to replace a society that yearns for excessive entertainment, intimacy, and material things. When the soul is revered and attended to, education becomes exciting and stimulating. Education refers not simply to enhancing the knowledge and skills of students, but to intellectual, religious, moral, physical, and all other aspects of the learner. True learning must affect behaviour so that the learner makes practical use of his or her knowledge. Education should nurture the soul and bridge the dualistic attitudes so prevalent in contemporary educational institutions. There is a need to connect the spiritual with the rational and to build bridges in the dualistic attitude prevalent in existing educational institutions [14]. The word "soul" bridges the dualistic separation of mind and body and promotes values, goodness, and beauty. "Education for all human beings" needs to explore in some depth a set of key human achievements captured in the venerable phrase "the true, the beautiful, and the good." [15].

Soul in education has been described by authors for a long time. Plato (430–347 BCE), the Greek philosopher, using the dialectic method, advocated an ideal world. His theory of education was related not simply to formal learning but to all aspects of one's life. The method he used required a teacher to ask leading questions for students to think about the meaning of life and truth. The purpose was to redirect students from a world of visual forms to an idealised reality and to stimulate them to define themselves through self-examination and analysis. Plato aimed at assisting people to live without violating the nature of their soul [16]. For Plato, the soul and spirit must be at the centre of all individual and communal relationships. This was a community meant to serve the people rather than to pursue pleasure, power, or wealth. Since then, scholars have written about those dimensions of life experiences that embrace depth, passion, relatedness, heart, and personal growth. They firmly believe that words like soul, spirituality, and spirit must form a part and parcel of education. To this end, the International Islamic University Malaysia (IIUM) has attempted to mainstream "spirituality" in the context of sustainability by giving deeper meaning to the quest of the future university [17]. It provides knowledge not simply from the utilitarian perspective but lays equal emphasis on physical, social, and spiritual development of the individual. It emphasises knowledge and skills without marginalising the spiritual and moral developments of young adults. Its curricula are geared towards enhancing the phenomenon of the global village.

4 International Concerns for Humanising Education

Prior to the 1990s, there was not much discussion on education with soul. Terms like soul and spirit were not considered appropriate for secular education. These terms, however, have gained relevance in contemporary times, including in education. Education should prepare students not only for a job and livelihood but also for life. Students should also develop their "moral, civic, and creative capacities to the fullest" [18]. Educational institutions are not meant just to teach a subject, but also to cultivate values and ideas and to live by those ideals. They must emphasise integrity, courage, self-sacrifice, and service to others.

These concerns have made educational institutions re-examine their vision and missions so as to make their institutions relevant and functional. A university should be truly international, emphasising interdependence between peoples and societies, understanding one's own and other's cultures while respecting pluralism. "Internationalisation" is suggested to be the key strategy to develop this kind of education. Equally, there is a need for universities to face the challenges related to development for sustainability. The relevance of sustainable human development, over time, deepened as the available natural resources were not able to keep pace with the increase in population. This was the major concern of the Millennium Development Goals (MDGs) and the UN approved the 2030 Agenda (SDGs) calling for comprehensive action to protect the planet, end poverty, and ensure peace and prosperity for all people. The focus is on "The World We Want" and to provide education based on UNESCO's Education for Sustainable Development focusing on five pillars of learning: to know, to do, to live together, to be, and to become [19]. Learning to know would let individuals benefit from educational opportunities that arise throughout life. Learning to do emphasises the acquisition of vocational skills through partnerships between the world of education and that of business and industry. Learning to be is to develop human potential to its fullest for a complete person. Learning to live together would require developing an understanding of others, their history, traditions, and spirituality [20]. Learning to become would equip students with tools and mindsets to shape the future of organisations, communities, and societies. The 2030 Agenda consists of 17 interdependent goals (SDGs) with 169 sub-targets or targets attached to the objectives. The themes are further elaborated on in the 5 Ps: people, planet, prosperity, peace, and partnership. Thus, the member countries of the UN agreed to humanise the natural world. The existing approach to higher education has brought about an imbalance between the planet, people, and prosperity with not much emphasis on peace and working together with others.

It is, therefore, essential to balance the 5Ps in the context of SDGs. This is advocated by UNESCO and supported by the International Association of Universities (IAU) in framing the 2030 Agenda for Sustainable Development. The IAU strongly advocates the potential use of technology as well as enhancing access and success to relevant knowledge and education for all [21]. Consequently, the IAU has been asking the authorities in higher educational institutions to include sustainable concepts and principles in their strategic planning, academic, and organisational work. The IAU would like to ensure that everyone benefits from quality education and lifelong learning.

5 Humanising Education

Clearly, education should be delivered to all equally without any consideration for geographical location, socio-economic status, and beliefs. The deadly novel coronavirus has brought this dimension to the fore. Based on its overall impact, it seems COVID-19 treats everyone equally without any regard for money or status.

COVID-19's emphasis on cleanliness and hygiene, use of face masks, physical distancing, and the like are meant for all. COVID-19 further highlights the need to be prudent and thrifty in expenses and consumption. "The more people are educated in adopting this lifestyle, the more people can share the limited available resources" [22].

COVID-19 is a reminder to respond to the shocks to the education system by seeking out new ways to deal with the learning crisis with determination. There is a need to replace, as shown in Fig. 1, the four Ms. model with four Hs to embrace the challenges of the twenty-first century [23]. This is a call to:

1. Broaden manpower for employment for humanity and sympathy, which is a feeling of pity and sorrow towards the misfortune of another. Sympathy is innate but it can be learned from childhood into adulthood to allow for the sharing of "feeling" beyond human-made barriers.
2. Couple Mind and invention with Heart and empathy, which is an ability to understand, acknowledge, and experience the feelings of another. Empathy is associated with "whole heart-fullness" and is part and parcel of most people's lives as the basis of inter-personal relationships. Empathy would make a person move out of his comfort zone, look for those who are suffering, and help them out of the sense of deep concern. This is where community engagement becomes important in the context of humanising education. Education must, therefore, inculcate this value of empathy among students so that they go out to the villages to "teach" and at the same time "co-learn" from members of the community, especially the youngsters who have no opportunity to have a relevant and sustainable education.
3. Expand machine-based Hi-Tech with Hi-Touch which is defined as "the ability to empathise with others, understand the subtle yearnings, and interactions of human beings, and pursue beyond every day for a new purpose and meaning...." (Pink, 2005: 17). It is based on compassion, which means to recognise the sufferings of others, relate to people in suffering and react appropriately to suffer-

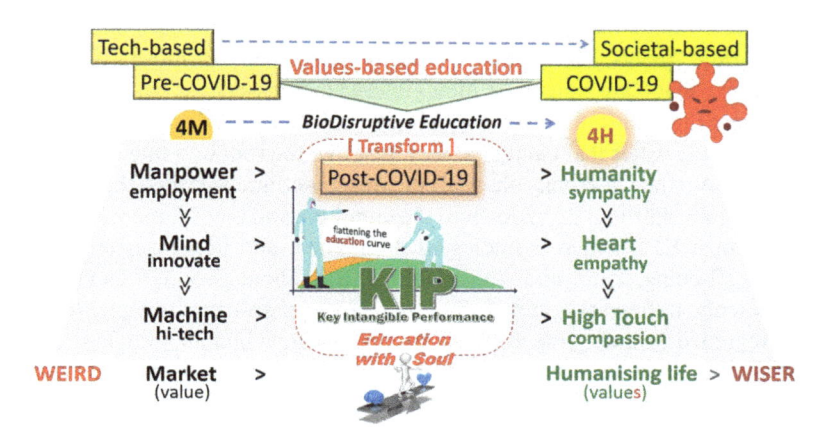

Fig. 1 Pre- and post-COVID-19 education models. (Source: [22])

ing, unconditionally. Compassionate people would help others, especially the needy, as part of their duty. They may be rich or poor but they share whatever they have. According to Mengzi, a Confucian philosopher, "If one is without the heart of compassion, one is not human" [24]. It is also said that "… we are entering the age animated by a new form of thinking. It is a new world where an aptitude for high concept and high touch is highly prized" [25]. The challenge is to humanise machines, especially robotics.

4. Replace Market (value) with Humanising (values), meaning valuing humanity in all interactions. Humanity has no price tag! Sympathy, empathy, compassion, and humanity are key attributes to living a healthy and dignified life in the context of *maqāṣid al-sharī'ah* (the objectives of Islamic law which is to create harmony with others and preserve public good) and *Sejahtera* (well-balanced and harmonious, which includes soundness, security, virtue, peace, and well-being). These values are apparently missing in education, which has become merely a tool to gain a sound and financially secure future as dubbed by education without a soul. Though financial security is a necessity of life, it cannot override the humanitarian considerations for a more just, equitable, and fair world. The ultimate purpose of education is to produce a complete, well-balanced person materially, intellectually, emotionally, and spiritually. Therefore, it is necessary to replace the utilitarian and mechanistic structure with an organic structure characterised by sympathy, empathy, compassion, and humanising values. This four Hs is values-based and would permit the all-round development of the whole and complete person.

6 The KPIs and KIP

It is also essential to re-look into the key performance indicators (KPIs). Although KPI is important, more important are non-intangible KIP (Key Intangible Performance), things that cannot be measured in the conventional and structured way. Intangible "assets" cannot be touched, felt, and seen. All those assets without a physical existence are known as intangible or tacit assets. At best, some people may see these intangibles but they may see about 10% of the things done. The other 90% is under the water that cannot be seen just like the root of a tree that cannot be seen. A dive into the water may show all the different dimensions of the intangibles, starting from the spiritual right down to the cognitive and environment and so forth. KPI, as against KPI, which is interested in numbers and figures, is about values, feelings, well-being, faith, and similar things. It is about goodwill and well-ness, beyond patents, trademarks, copyrights, and computer software, which are conventionally regarded as intangible assets. It is about being a human who acts to benefit oneself as well others in order to please the Lord of the universe. This makes KIP a very noble idea that is closely linked to the essentials of *maqāṣid al-sharī'ah* producing a person (IIUM Roadmap, 2019–2020) who is well-balanced (*insan sejahtera*). Stated differently, KIP is about internal (microcosmic) values and not simply concerned with external (macrocosmic) values. It is an inside-out process. To be sure, KPI is important, but equally important is KIP and the two need to be well-balanced and harmonised holistically synergised by the combination of the head (cognitive), heart (socio-emotion), and hand (skills). Working to fulfil the requirements of KPI to the neglect of KIP will not result in a well-balanced healthy state of mind. The reverse may be true, however. Working on KIP may lead one to achieve the requirements of KPI as a true servant of God, the Almighty.

7 Sejahtera and the Global Campus

To bring back the soul, education with soul, is to humanise education, to develop a complete person's potential in a holistic and integrated manner to produce a balanced and harmonious individual, *insan sejahtera*, [26] laced with intellectual, spiritual, emotional, physical elements with firm faith in and devotion to God, the Almighty. This idea of achieving a quality of life and balance is very well captured in an indigenous Malay concept, *sejahtera* [11] which is the fountain-head of good values or virtues that will lead to a balanced human person, *insan sejahtera*. This will help achieve a peaceful, fair, and just global community framed on ethical and moral principles in tandem with the aspirations of the United Nation's SDGs.

The higher educational institutions must not simply be "international" but "global" in the real meaning of the word. The terms global and international, though used interchangeably, are different in scope and approach. The term "international" refers to issues and concerns of and interaction between two or more countries irrespective of national boundaries. Compared to the term global, which is all-encompassing and worldwide, the term international has a smaller scope encompassing only a few countries. The word "global" refers to the entire world and not just one or two regions. It is worldwide, universal, unlimited, unbounded, general, and comprehensive. It is "macrocosmic." It means an integration of the various countries of the world to help socio-economic development through collaboration and cooperation rather than competition to be ranked and counted. This would lead to increased economic, political, and socio-cultural interactions, and sharing between the various partners (nations) in a more equitable and just way. Global means the integration of the people of different regions (including indigenous communities) as a single unit who join together to resolve common issues that concern and affect the planet as a whole. The global approach is holistic, interdependent, interconnected, sustainable, and balanced. The overarching 5Ps (People, Planet, Prosperity, Peace, and Partnership) targets, as envisaged by SDGs, speak the same language. It is about protecting and promoting life, intellect, progeny, resources, and a moderate way of life.

This requires a re-thinking of education to foster the competencies and values that people, societies, and economies need in harmonious and balanced ways. Education is the key to the global integrated framework of SDGs intertwined with *maqāṣid al-sharī'ah* and Sejahtera [26]. This calls for a conducive learning environment and for adopting new approaches to learning. It means learning and adjusting different areas of one's life towards sustainability, through re-training, deeper reflecting, observing, and co-learning globally. It embraces a larger global agenda ranging from climate crises to massive migratory exodus and the many injustices exposed by the pandemic affecting mental health and socio-emotional challenges.

Universities must establish global campuses essential to achieve the United Nations' SDGs, especially the fourth goal dealing with quality education and lifelong learning for all. The global campus must be conducive health-wise in ensuring sustainability as an overarching objective. Essentially, it is about making the campus stress-free. This necessarily means it is drug-free, including tobacco-free and smoke-free. In such a campus, the faculty shares with others personally held (tacit) knowledge areas, interests, and passions. They develop diverse teaching methods and techniques to facilitate successful learning experiences for all involved. They encourage students to appreciate their own learning styles in a celebratory, stress-free academic atmosphere. The administrative staff experiences the opportunity to influence the organisational direction and curricular activities of their campus.

A global campus values diversity and multiculturalism in its students, faculty, and administrative staff. Diversity is not merely about race and gender but includes religion, age, and socio-economic status, not forgetting geopolitics, norms, and cultures. Such a diverse campus challenges predisposed stereotypes or norms that may have been developed during adolescence. Campus diversity helps students improve their affective and cognitive skills, such as resiliency, critical thinking, and problem-solving. It fosters mutual respect and teamwork and produces graduates who are capable of working as good citizens in an increasingly complex, pluralistic world. Some universities and higher institutions have formal diversity programmes to help all students accept, understand, and celebrate the differences.

To create a peaceful, fair, and just global community, the university has to create a harmoniously balanced community and create its own environment, its own institution, its own research, and its own global relevance. The university community must be fully familiar with the elements that make the university. To be institutionally ready, the university community must have similar ideas and similar interpretations to make the university truly international, global, and sustainable.

Institutional readiness means that members of the University community are determined and able to implement a change. They are disciplined and are capable of working across teams, by breaking the silos. This will lead to humanising education characterised by intellectual honesty, integrity, and hard work. This, in turn, will produce a better human being bringing about peace, and well-being in a sustainable and balanced way.

Interestingly, COVID-19 also highlights the ideas of wellness and quality. It teaches cleanliness, hygiene, and "physical" distancing and avoids waste or extravagance. These are the ways to ensure equality so that people can share the limited available resources. Finally, COVID-19 teaches that people must stay sustainably productive under all circumstances and work from home or anywhere. It teaches to converge on values, integrity, and discipline as well as be grateful and giving back where and when appropriate, manifesting *raḥmatan li'l 'ālamīn* (mercy to all the worlds).

This means there is an urgent need to reconstruct a new academic framework to fit into the post-pandemic challenges ahead in flattening the education curve as it were. In this regard, the International Islamic University Malaysia (IIUM) has co-created the Sejahtera Academic Framework that emphasises empowerment, flexibility, innovation, and accountability in dealing with the ambiguous times ahead [27].

8 Concluding Thoughts

There is much to be learned in the traditions that encourage an education emphasising not only the mind but also the soul. Most contemporary educational institutions are characterised by the loss of soul in the lives of students as well as teachers. This can be seen in their expressions of daily living and their lack of resilience. This is evident in the growing disconnection between teachers and students. Teachers complain about the reluctance of students to engage in deep learning experiences, and students are unhappy about teachers not being interested in their welfare. Learning has become a matter of skilful application of policies and strategies that deliver grades and degrees. Teachers spend their energies on applying the latest strategies (including technologies), and students focus much of their time and energy on earning a degree to get a good job. Consequently, there is a loss of the soul. Symptoms of loss of soul can be seen in acts of violence and addictions, and in distorted ways students respond to daily life.

There is, therefore, a need for a coherent, thoughtful, and accessible response to the question of how to bring back the soul into education with a higher sense of purpose. Such a response must be grounded in wisdom, and an ability to integrate theory and practice. This is what the International Islamic University Malaysia (IIUM) has been aspiring to achieve.

It is essential to bring about a balance between the utilitarian and mechanistic structure aimed at employment, innovation, high-tech and market value, and a structure characterised by sympathy, empathy, compassion, and values that humanise life. Once there is a balance between machine and mind, educational institutions will change radically. Likewise, the key performance indicators (KPIs) must be rightly balanced with KIP (Key Intangible Performance), things that cannot be measured in the conventional and structured way. Admittedly, KPI is important, but equally important is KIP and the two need to be well-balanced and harmonised holistically synergised by the combination of the head, heart, and hands.

Education must be based on respect and partnership. It must take into consideration the social, economic, spiritual, and environmental dimensions. It means learning, and adjusting different areas of one's life towards sustainability, through re-training, deeper reflecting, observing, and co-learning globally. Universities must establish global campuses to ensure sustainability as an overarching objective. In such a global campus, the faculty shares with others their personally held (tacit) knowledge areas; students appreciate their own learning styles in a celebratory stress-free academic atmosphere; and the administrative staff experiences the opportunity to influence the organisational direction and curricular activities of their campus. There are many, overlapping, and cohesive interactions among all members of the organisation. They value diversity and multiculturalism culminating in humanising education characterised by intellectual honesty, integrity, humility, and hard work.

In the post-COVID-19 world, teachers and scholars are needed to uphold integrity. They will explain and practice ideas of readjustments for society. They will articulate the ways and means of reconfiguring the "new normal" and will no longer be held ransom to imperial businesses. Students will be guided to be compromising, to be kind, and to accept diversity. COVID-19 shows clearly the ability of the people to cooperate with compassion, amidst catastrophe. Teachers will contribute to factual and intelligent dialogue and project truth as the new pandemic. They will consistently choose to do the right rather than the expedient thing and to be values – rather than needs-driven. It is instructive to learn how to best navigate between the similarities and differences of the COVID-19 and educating for the future framed by the SDG.

In sum, existing educational understanding has strayed far from the foundational ideals of truth, balance, beauty, and goodness. The soul is often ignored and hence learning is dull, disconnected, and trivialised. A soul culture has to be built on attitude and trust. To introduce the soul into education, to humanise education, is a gentle enterprise but it requires deep-seated strength and security. There can be nothing more important to the civilisation than to bring the soul back into education as the interaction between human beings and advanced technology share the global stage for the better of humanity and for the mercy to all.

References

1. Abdul Razak, D. (2021, March 5). *Concept note, UNESCO experts workshop. Futures of Higher Education*. UNESCO: IESALC.
2. Altbach, P. G., & Reisberg, L. (2018). Global trends and future uncertainties. *Change: The Magazine of Higher Learning, 50*(3–4), 63–67.
3. Collini, S. (2012). *What are universities for?* (p. 98). Penguin.
4. Brown, W. (2011). Neoliberalized knowledge. *History of the Present, 1*(1), 113–129.
5. Smyth, J. (2018). *The toxic university*. Palgrave Macmillan.
6. Abdul Razak, D. et al. (2020, May 20). Time to free our universities. *The Sun Daily*.
7. Schwab, K. (2016). *The fourth industrial revolution: What it means, how to respond*. https://www.weforum.org/agenda/2016/01/the-fourth-industrial-revolution-whatit-means-and-how-to-respond/. Accessed 14 Aug 2017.
8. Abdul Razak, D. (2018). *Fourth industrial revolution: The leadership dilemma*. USIM Press.
9. Fitzsimons, P. (2002). Neoliberalism and education: The autonomous chooser. *Radical Pedagogy*. http://radicalpedagogy.icaap.org/content/issue4_2/04_fitzsimons.html. Accessed August 17, 2019 [Google Scholar].

10. Seldon, A., & Abidoye, O. (2018). *The fourth education revolution: Will artificial intelligence liberate or infantilise humanity* (p. 22). The University of Buckingham Press.
11. Abdul Razak, D. (2019). *Leading the way*. IIUM Press.
12. Abdul Razak, D. (2013). University in times of civilizational transformation: Challenges and responsibilities. In *Civilizations and Higher Education – In search of Great Learning* (pp. 35–51). Kyung Hee University.
13. Lewis, H. R. (2006). Excellence without a soul: How a Great University forgot education. *New York: Public Affairs, 2006*, 2–3.
14. Moore, T. (1992). *Care of the soul* (pp. xi–xxii). Harper Collins.
15. Gardner, H. (1999). *The disciplined mind: What all students should understand* (p. 18). Simon & Schuster.
16. Lodge, R. C. (1970). *Plato's theory of education* (p. 198). Russell & Russell.
17. Moten, A. R. (Ed.). (2020). *Spirituality and sustainability: Experiences of the International Islamic University Malaysia*. IIUM Press.
18. Fish, S. (2008). *Save the world on your own time* (p. 11). Oxford University Press.
19. United Nations. (2019). *About the sustainable development goals*. https://www.un.org/sustainabledevelopment/sustainable-development-goals/. Accessed 25 Feb 2021.
20. UNESCO. (2014). *Shaping the future we want. UN decade of education for sustainable development (2005–2014)* (Final Report, pp. 93). http://unesdoc.unesco.org/images/0023/002301/230171e.pdf. Accessed 15 Feb 2021.
21. International Association of Universities. (2018). *Digital transformation of Higher Education*. https://www.iau-aiu.net/technology. Accessed 20 Feb 2021.
22. Abdul Razak, D. (2020). *Essay on Sejahtera – Concept, principles and practices*. IIUM Press.
23. Abdul Razak, D. (2021). *Leading for Sejahtera – Humanising education*. IIUM Press.
24. Choi, D. (2019). The Heart of compassion in *Mengzi* 2A6. *Dao, 18*, 59–76. https://doi.org/10.1007/s11712-018-9641-7. Accessed 2 Mar 2021.
25. Pink, D. H. (2005). *A whole new mind: Moving from the information age to the conceptual age*. Riverhead Books.
26. IIUM. (2019). *IIUM roadmap 2019–2020: Whole-institution transformation*. IIUM: Office for Strategy and Institutional Change.
27. IIUM. (2021). *Sejahtera academic framework*. IIUM: Office of Knowledge For Change and Advancement.

Social Science and Humanities in Future University

Mahjoob Zweiri and Lakshmi Venugopal Menon

1 Introduction

Social sciences and humanities have made unparalleled contributions to understanding major milestones of world history and modern times. The world wars, many other wars, colonization, French revolution, changes in Europe, industrial revolution, the great migrations – all have been studied and interpreted by social scientists and humanities practitioners. Pure science and mathematics cannot address such complex social phenomena. Social studies help one make sense of the world around them. It enables individuals to participate in societies by providing one with political and social skills, understanding about history and culture, and encourages critical thinking. Purposeful social studies are integrative, meaningful, challenging, value-based, and active. However, since 2009, public and private funding for arts and humanities has reduced the world over. This decline was not a new phenomenon, and once again, the propagation of interdisciplinary, multi-disciplinary, cross-disciplinary, and trans-disciplinary fields was identified as methods to safeguard and maintain humanities and social sciences.

Moreover, there needs to be a systematic roadmap to the Universities of the Future. One can envision Future Universities which encourage more interdisciplinary fields that are sensitive to the job market and offer greater job opportunities. Emphasis on vocation-alization of education and student-centered education will also be unique features of Future Universities.

The twentieth and twenty-first centuries posed many challenges to universities. Radical disruption, driven by globalization, a new work order, and fleeting technological advancements, is forcing universities to redefine and reassign their roles and what they offer to students. The crux being the age-old debate of whether the field

---

M. Zweiri (✉) · L. V. Menon
Gulf Studies Center, Qatar University, Doha, Qatar
e-mail: mzweiri@qu.edu.qa

© The Author(s), under exclusive license to Springer Nature Switzerland AG 2023
M. Ali S A Al-Maadeed et al. (eds.), *The Sustainable University of the Future*,
https://doi.org/10.1007/978-3-031-20186-8_8

of science must concentrate on pure sciences and disassociate from social sciences. A debate common in the academia and, many at times, the global polity. An appraisal of this debate in the twenty-first century would require a review of the demands, needs, opportunities offered, and challenges faced by social sciences and humanities. It also calls for addressing the general perception of science being pure science and social science being "all-talk."

This attitude had caused the shrinking of the fields of social sciences and humanities in the past. Various departments were merged together. Some even shut down. For instance, United States witnessed sociology and philosophy departments being merged or closed as a result of policy shifts and consequent lack of public and private funding. This led to a different strategy to safeguard and maintain social sciences and humanities – the emergence of interdisciplinary, multi-disciplinary, cross-disciplinary, and trans-disciplinary fields such as gender studies and area studies. This new approach was effective.

This research contribution is not a case study of a particular educational institution. It does not focus on a particular region. The paper looks into the debate regarding social sciences and humanities in universities. Through the appraisal of existing arguments, contentions, and challenges, the paper will study how social sciences can survive the lull that is being experienced in academia. The main issues highlighted in the course of the paper are the challenge of image faced by social sciences, the issue of job opportunities, and the actual impact experienced and exerted by social sciences.

The paper commences with the section "Social Science and Humanities in today's University" – an analysis of the current situation of social sciences and humanities by appraising them in the context of current universities. The role of state perception is discussed here. In the next section "What is Future University?", the authors will provide a description and conceptual framework for Future University or universities of the future. The third section "The Future of Social Science: Attempts to Survive" will analyze the existing arguments, contentions, and challenges faced by social sciences and humanities in the academic world. The paper will conclude with the section "The Future: What lies ahead?". This section will look into the future of social sciences and humanities in the Future University. It will also briefly touch upon the importance of social sciences and its undying appeal.

This paper aims to address certain key questions. Why particular degrees are needed? Are there ample job opportunities that have job profiles which match the knowledge and skills acquired from the degrees? To what extent do the existing degrees serve larger purposes, such as contributing to the national development? The main question that the paper addresses is whether the Future Universities would continue on the same track of their predecessors. Will they also focus on the above-mentioned questions which seem to be of concern to the current universities or would they prefer to go beyond these current questions? Would Future Universities go beyond knowledge and skills? Would they go beyond the general perception of degree being a means to a job opportunity? How can social sciences and humanities survive the current lull it is experiencing and transform to a stage wherein it can exert a real presence and impact in the Future University?

In this light, another issue requires attention. The change is a general state perception that higher education is the citizens' right and the government has the responsibility to invest in and offer its citizens higher education opportunities through state universities or private universities. This perception of higher education being a citizen's right has changed in many parts of the world. Will this shift in perception continue in the future? Would future governments opine that education is not a state responsibility and urge its citizens to invest in themselves if they wanted. With more countries discounting higher education as a right, it is a crucial aspect one must consider in the study of Future University. State perception will substantially affect the projections, prospects, and direction of Future Universities. In the beginning of 2020, a senior adviser to the UK government Dominic Cummings called on mathematicians, data scientists, and physicists to join him in the government. He, however, left out practitioners of humanities and social sciences. Complex societal challenges such as issues concerning the environment, migration, national and geographical identity, inequality, and poverty cannot be solely solved by mathematics and science [20].

At this juncture, it would be beneficial to take note of the contributions of sociology, political theory, and anthropology as fields of sciences toward the wider appreciation and acknowledgement of the impact of social sciences. Anthropology alone has played a crucial role particularly in the period of the COVID-19 pandemic. It can be viewed as the ultimate anthropological experiment that large societies were part of. During this pandemic, vast volume of research was done on online learning mechanisms and practices. These included studies on the psychological and emotional effects of school disruption on children, teachers, parents, and institutions at large. These studies threw light on the significant role of social sciences outside the realm of academia and universities – in the real world. It showed how important social sciences is in making sense of the world around us. One can argue that it also, in a way, enhanced the image of social sciences within the academia, thus making a stronger case for the preservation and development of social sciences in universities.

Similarly, political theory has also always made a strong case for the enhancement of the fields of social sciences. Politics remains an indispensable facet of the international community.

2 Social Science in Today's University – Criticisms and Challenges

The new interdisciplinary approach also did suffer at various points. Pinar Bilgin, in his article titled "Is the 'Orientalist' past the future of Middle East studies?", mentions "the relevance debate" that gained momentum during the end of the Cold War

[5]. At that time, the sponsors of Area studies (an interdisciplinary field of study) such as Social Science Research Council (SSRC) and the American Council of Learned Societies (ACLS) dissolved their joint Area Studies committees and withdrew funds. Further, they channeled their research toward more relevant post-war concerns such as effects of structural adjustment programs and post-communism transition [15]. However, Ford Foundation continued to support Area studies and provided space and resources for its reshaping. Within the Cold War context, the crisis of Area studies, which Bilgin explains as "a clash between 'theory-free' Area Studies and 'scientific' approaches to world politics" also features [21].

Bilgin also talks about the division of labor and hierarchy between Area Studies specialists and disciplinary generalists. This is a tussle that still exists in this field, hence, rightfully mentioned by Bilgin [5]. This example throws light upon the manner in which politics and state perception impact the fields of social sciences and humanities. Kramer argued that area specialists must drop interest in "fashionable theory" (which he associated with Edward Said's "Orientalism") and look critically inwards to create more straight-forward policy-relevant work. Kramer said they must adopt approaches that would "explain and predict" developments and change in the Middle East while returning to the roots of Oriental studies, which according to him would "restore some continuity with the great tradition" [3]. Separately, social sciences could move toward a disciplinary-orientated view that includes formal theory, statistical methods, and mathematics. Social sciences, after all, do produce descriptive work which leads to diverse empirical data that require incorporation by universal theories in disciplinary methodologies.

Nevertheless, one cannot ignore or discredit the need for disciplines. For instance, area studies foster close links with disciplines. Globalization, the rise of rational choice theory in the social science and the rising popularity of the cultural and postcolonial studies, has in fact increased the need for area-based knowledge. The need for the latter cannot be dismissed in an increasingly globalizing world [15]. Area studies must utilize disciplinary knowledge and methods to further localized or contextual knowledge that is necessary in the modern times [3].

At this juncture, it is crucial to analyze the reasons behind the trend of decline in social sciences and humanities. Some have attributed the decline to two major developments: the political criticism of the humanities and the funding cuts to literature, history, and arts programs at public universities. In the United States, Republican governments have proposed to rebalance funding toward "practical" subjects by cutting down on funding to humanities departments at state universities.

In 2013, North Carolina's governor Patrick McCrory expressed his intention to change the state's legislation on higher education funding to ensure that the decision is not dependent on the demand for the course but on the job opportunities offered by the course [10]. Like other critics of social sciences and humanities, McCrory did not want the taxpayers' money to subsidize subjects or courses that did not lead directly to job securement by students. In Pennsylvania, some universities planned to completely shut down music and language departments [16]. Meanwhile, the debate regarding massive budget cuts to the National Endowment for the Humanities continues across the United States. In 2015, Japan witnessed massive budget cuts to and shutting down of humanities and social sciences [6, 8].

Since 2009, across the world, public and private funding for arts and humanities has reduced. This decline was not a new phenomenon. Between 1970 and 1985, humanities enrollments dipped from 17.2% to around 7%. A less direct reason is that there has been a substantial shift in the subjects chosen by women. The latter no longer shy away from pure science subjects and courses that give them promising professional careers. They choose medicine, finance, and engineering over liberal arts, education, and history. There was a favoring of pre-professional degrees. This is a deeper story that has been masked by the histrionics [25].

What made women turn to other subjects? What will be the implications of their choices? This could have been due to increased equality, need to enter the workforce, for higher pay, or to increase their desirability over male candidates. The reason for such a shift in choice of education and subject still lacks a concrete answer. The impact of feminist movements and ideology must not be undermined in this respect. This is in fact is a topic that could be expanded into an independent research.

3 What Is Future University?

Today, the humankind is witnessing the onset of widespread technological revolution, of unprecedented scope, scale, and complexities, that will fundamentally change the norms of work and life. In order the ride this wave, a comprehensive, integrated, and holistic approach that brings together the civil society, academia, public, and private sectors is necessary. The first Industrial Revolution mechanized production using steam and water power. The second created mass production using electric power. The third automated production using electronics and information technology. The current or Fourth Industrial Revolution, Klaus Schwab the founder and executive Chairman of the World Economic Forum, argues is building on the third revolution's digital prowess. He opines that it has been occurring since the mid-twentieth century.

The characteristic of this revolution is the "fusion of technologies that is blurring the lines between the physical, digital, and biological spheres." The current revolution is different from the former in terms of scope, velocity, and systems impact. It is growing in an exponential pace, rather than a linear one. Moreover, these developments have no historical precedent. Furthermore, it is causing disruptions world over. The effects have even been visible in the education sectors globally. Billions of people being connected with each other digitally has led to unprecedented levels of storage capacity, opportunities, and access to knowledge. This coupled with the pandemic has caused significant disruptions in the education sector – particularly, one may argue, in the higher education sector. It has brought down some barriers to higher education. The emerging technologies of robotics, artificial intelligence, 3-D printing, computing power, availability of vast amounts of data, novel algorithms, and others are making a strong case for virtual educational platforms. Travel restrictions and social distancing policies of the COVID-19 pandemic catalyzed this shift. In fact, some institutions are now offering purely online educational programs for students [18].

In recent times, pure sciences have dominated over social sciences and humanities. Social sciences are facing budget cuts and rollbacks in funding. Meanwhile, technological advancements are also pure science driven. This leads to shortfalls and datedness in social science research. Gus Wachbrit identifies poor citation practices, navigation troubles, the question of sustainability of open sources, lack of peer-review system for tools, the role of big technology and media bigwigs, and innovation as immediate challenges faced by social science's new technology [26]. This paper will attempt to study how social science and humanities will be perceived in Future University. This requires conceptualization of Future University. Future University is one that does not rely on pure knowledge but incorporates vocational skills, and basic and advanced technology. Future Universities seem likely to approach social sciences and humanities by maintaining interdisciplinary courses and pushing for the larger acceptance of this approach. In this pursuit, Future Universities would encourage more interdisciplinary fields that are sensitive to the job market and offer greater job opportunities.

The aim of a Future University would be the generation of a paradigmatic change in the manner in which higher education institutions, industries, businesses, and private and public authorities cooperate within the currently transforming framework of the global market and educational industries.

Future Universities are expected to enhance the lifelong learning experience of individuals. Certain attributes could be associated with Future University: first, on-demand, customized education. The courses would be tailormade and suit the specific job profiles and career interests. Second, there would be an amalgam of various degrees and would likely offer courses with shorter periods for completion and qualification. Third, the universities would provide quality state-of-the-art career advice and career-management support. Last, there would be dedicated teams in place for collaboration and promoting entrepreneurship. There could also be four scenarios that Future Universities would find themselves in: championing, commercializing, virtualizing, or disrupting. Future Universities would most likely be precincts of innovation that focus on distance education by converging formal and informal teaching and learning methods. Emphasis on vocation-alization of education and student-centered education will be prevalent.

As a result, interest in and prevalence of preparation schools such as the British school "Doha college" could increase. These Future Universities would encourage the application of research for positive community impact. Moreover, there would be zones wherein universities and industries are co-located for the ease of collaborate on projects. These future educational institutions would encourage moving between borders to gain a global outlook. They would encourage technical and transferable skills. Student-focused e-learning platforms with state-of-the-art technology that necessitate immersive learning will be developed and implemented. It would be a true blending of traditional and non-traditional teaching and learning methods.

But there needs to be a systematic roadmap to the Future University. "Universities of the Future," a Knowledge Alliance Project co-funded by the Erasmus+ Programme of the European Union, is a pioneering project that lays down a five-step roadmap for the university of the future. First, the laying of the foundations. In order to produce a comprehensive plan, detailed analysis of the readiness and status of maturity of regions and industries, and its impact on skills of local force is necessary. Equally important is studying the potential of integration of higher education and the job markets. It aims to benchmark and showcase best global practices which support regional-driven re-industrialization processes and to form a blueprint for cooperation among stakeholders to "accelerate innovative practices within the fourth industrial revolution." It is important for developing a blueprint for innovative practices in educational sector [18, 19].

Second, co-creating a collaborative strategy by associating with key players. This will help boost engagement of stakeholders from an early stage. It will also nurture joint experiences which are built-on on open partnerships. Third, to co-develop, co-test, and co-verify educational resources and reforms that are feasible. This will help develop and pilot innovative lessons and equip students with the valorized skills. It will aid continuous training programs regarding lifelong learning methods. Additionally, developing and piloting a post-graduation course dedicated to creative strategies of Industry 4.0. will upskill competences.

Fourth, identifying valuable resources and tools. This step looks into developing and testing resources which target higher education institutions, government decision-making bodies, and businesses "to promote the creation/updating of strategic plans and definition/implementation of specific actions toward industry 4.0 and education 4.0" [18]. It also targets educators and trainers in order to encourage them to use innovative methods and technologies in the pedagogical approaches. Last, designing and building a virtual teaching and learning apparatus. This will set up an online hub which fosters direct contact between all relevant players and provide a platform for discussion, deliberation, and debate.

Understanding the changing role of degree in the job market is important. A degree's relevance is decreasing. Instead of hiring a specialist, companies are choosing to hire a fresher and fine-tune them as per the job requirements. Many companies found that the newly hired graduates were lacking required skillsets for performing effectively. Thus, companies, such as Google and Tata Consultancy Services, commenced crash training programs for the newly hired freshers. To bridge this gap, current universities are attempting to match degrees, specializations and job opportunities; in other words, justifying the need for a degree with respect to existing job opportunities. The market tendency is reflected in the universities' focus on primarily skill-based courses. Universities have moved away from knowledge-based courses to skill-based courses, giving way to vocational training.

4 The Future of Social Sciences: Attempts to Survive and the Impact of COVID-19

In Africa and the Caribbean, humanities are being viewed as catalysts for economic development [7]. Such attitudes will also help in the preservation and upliftment of related fields of education. During a conference organized and report produced by Centre for Education Policy Development (CEPD), in the Republic of South Africa, Dr. Sarah Mosoetsa, Deputy Director of the Ministerial Task Team, stated that the revitalizing of Higher Education in South Africa had to do with the need to recognize the current institutional context, history, society, and the wider social science context (domestic and international). This shows that the focus on STEM (Science, Technology, Engineering, and Mathematics) or pure science in Universities is not necessarily of universal nature. The attitude is not heterogenous globally. One may wonder if interdisciplinary studies are more common as a strategy to "save" the Humanities and Social Sciences in developing economies? The answer to this, however, is a research project by itself and will not be explored in the course of this chapter.

The global COVID-19 pandemic has posed multiple challenges to the existing world order. Among world trade, world trade, health care, human security, and travel, education systems are also in crisis. Kalinga Tudor Sliva argues that the pandemic could be a blessing in disguise for social sciences as it challenges the current modus operandi of social sciences and urges social science researchers to revamp existing theories and approaches regarding human coexistence with living entities including microbes in a dynamic world; one with shrinking resources and swelling plethora of economic, social, and ecological challenges. He is of the opinion that social sciences need to self-reflect regarding the nature of knowledge production and addressing challenges posed by the pandemic for research methods, ethics, and epistemological framework. He rightly argues that social science will play a crucial role in understanding the new world order that will come about due to the pandemic. COVID-19's "impact on life, livelihoods, economic processes, social support mechanisms, vulnerable groups in society, social harmony and international relations" are cited as areas of direct relevance to social sciences [22].

Silva says that COVID-19 poses methodological and analytical challenges. "The aim of methodological research in the social sciences is to learn what conclusions can and cannot be drawn given empirically relevant combinations of assumptions and data" [13]. As Wood, Rogers, Sivaramakrishnan, and Almeling argued in a recent article "Resuming field research in pandemic times" reported in SSRC website, existing intensive field research methods may be unfeasible due to the following factors: first, the uncertainty of the course, duration, and consequence of the pandemic; second, turmoil faced by potential social science subjects; third, inability to perform social research techniques that require close proximity with subjects such as interviews; and last, ensuring the researcher will not be affected by the virus. Analytical challenges stem from the fact that hygienic and conceptual tools of response to the pandemic were evolved in Western countries that have a particular background in sanitation and science and technology. Thus, ideas of quarantine, social distancing, and respiratory hygiene would not have a uniform mean across countries and cultures [27].

Culture-specific social cleavages, prejudices, and agitations must also be factored in while assessing new challenges for social analysis.

COVID-19 upended the higher-education world. During its onset, students were abruptly sent home and classes were moved online [23, 24]. Administrators of educational institutions scrambled to put a plan in place. The financial challenges were severe leading to bankruptcies, business failures, and more need for state bailouts. The pandemic pushed private and public educational institutions and colleges to the breaking point [23]. Expenses climbed while revenues dropped significantly. Learning was moved from the classrooms to the Internet. Countries across the world went into an experimentation mode of sorts. In this manner, COVID-19 accelerated the trend toward online education [14]. It also raised fundamental questions regarding a college degree's value. Students and parents began demanding refunds for their tuition fees [4]. This made academics and policymakers alike wonder what the lasting legacy of COVID-19 would be on higher education. Would the pandemic lead to college closures? Is the online learning a temporary phenomenon or a permanent transformation?

In 2018–2019, higher education futurist Bryan Alexander wrote about the possibility of a major global pandemic transforming the education industry. His views are chillingly prescient. He forecasts a "toggle term" scenario wherein the pandemic continues and regional flareups become commonplace. Such a scenario would lead to a hybrid learning system with online and in-person teaching and learning [2].

In reviewing the Future University and the paradigmatic changes that are required post-pandemic, the new societal demands of employees that require a stronger combination of "soft skills" traditionally found in the subjects taught in humanities and social sciences make a strong case for the increasing relevance of the fields. Proficiency in soft skills such as empathy, team building, communication, team work, and leadership is being sought out by multi-national corporations, world over. Companies no longer simply look for professional qualified engineers. They prefer professionally qualified individuals who possess the preferred soft skills. The learning, teaching, and enhancement of such social skills is in the forte of social sciences. Thus, such preferences in the job market will give new life to the social science and humanities disciplines in the Future Universities. The Reskilling Revolution that was launched at the World Economic Forum's 50th Annual Meeting aims to provide one billion people with better education, skills, and jobs by 2030. It also announced a novel alliance SkillsLink which is expected to be a crucial factor in the Reskilling Revolution. It focusses on "making skills the currency of the labour market." It aims to recognize skills-based credentials, adopt skills-based workforce policies and strategies, adopt a common language for skills, and partner for skills-based learning delivery systems. The initiative has brought together 18 multinational companies, online-learning providers and businesses, technological leaders, and education ministry personnel. Together, they aim to create a foundation for a new 4.0 Education system which features as a key part of economic recovery policies [28].

Ultimately, what impact will the pandemic have on social sciences? Will the new challenges and developments generate renewed appreciation for social sciences or damage its relevance citing the inconsistencies between the existing disciplines and theories that compete with each other? Will social sciences and humanities emerge as the preferred pathway to tackle the novel global challenges? Nevertheless, the research agenda of social scientists will undergo profound changes.

Online education is here to stay. Its relevance and prevalence will only grow with time. Meanwhile, a survey by survey by the Arab Council for the Social Sciences shows that the global pandemic has had particularly negative consequences for the practitioners of social sciences and humanities in the Arab region. Rollbacks on funding, travel restrictions, and social distancing are causing hurdles for social science research, particularly for anthropologists [9]. In-person field researches such as ethnography are likely to be among the last scholarships to return to normalcy post the pandemic. The variance and fluctuations in political, travel, health, and social policies and conditions significantly impact ethnographers. In the present scenario, ethically and logistically feasible research is difficult. In short, it is not researcher-friendly.

The impending crisis faced by the social sciences calls for inevitable actions such as perception change and contribution to the real life using various languages and existing technology. These are attitudinal changes with respect to the practices that social science researchers are accustomed to. Social scientists must address the pressing challenges of today through research, dissemination, and policy analysis.

Ronald Pohoryles argues that the future of the social sciences is dependent upon reverting to the classics. Although his argument seems to be facing contradictions from complexities of economy, society, and policies, the field of social sciences is experiencing greater levels of specialization, via national policies, traditions, and new schools of thoughts. Ronald argues that thinkers who developed classical literature such as Max Weber, Karl Marx, or Adam Smith were faced by much fewer complex societies. Social science's mission is to "contribute knowledge to societal development." Simply producing "general theories" will not suffice in answering the problems of contemporary complex societies. He also offers pragmatist theories based on "patchwork theories" as a solution for this dilemma.

One can argue that research programming and social science research are two sides of the same coin. Research requires funding. The latter comes with legitimate expectations from the funders – both in the private and in public sectors.

Nevertheless, there are significant differences between research programming and social science research in terms of priorities, perspectives, and expectations. This affects the autonomy and creativity of the research community, thus causing greater fears of not having a "successful" outcome. Innovation could lead to success or failure, as it is the designing of new actions and/or structures. Meanwhile, commitment, creativity, imagination, and risk taking are also essential [11]. Complex societal challenges make mere technological solutions inadequate. The importance of context and non-scientific knowledge must not be ignored. The challenge of social sciences and humanities is to define new research strategies and a new epistemology.

The propagation of interdisciplinary, multi-disciplinary, cross-disciplinary, and trans-disciplinary fields has been considered as a method to safeguard and maintain humanities and social sciences. However, it must be understood that universities and societies do not appraise social sciences and humanities in the same manner. Attitudes, perspectives, and perceptions regarding the purposes of social sciences and humanities are different.

Interdisciplinary research can be defined as either the research that occurs across multiple existing academic disciplines, or the research that occurs in the space between disciplines. Interdisciplinary research can be conducted by a single researcher working across various disciplines ("single scholar interdisciplinary research"), by groups of people from different disciplines who are work within their own disciplines but collaborate on specific projects ("multidisciplinary research"), or by research groups that work across disciplines ("interdisciplinary teams").

Interdisciplinary research possesses certain benefits: First, it presents a more thorough research approach to challenges as opposed to pure disciplinary research. Many pressing issues of today that are faced by the society cannot be solved by traditional disciplinary research approaches. Interdisciplinary research is better suited to address these problems as they address common issues from multiple perspectives. Second, the cross-fertilization of research methodologies through new developments in data availability. Social media data analysis in political science and fMRI in economics are examples of such novel developments. Such new methodological techniques help better tackle the issues on ground. Multidisciplinary research will provide a more holistic addressing of pressing challenges faced by contemporary societies. Third, the establishment of unidentified new fields of study. There is a need for establishment of novel unexplored interdisciplinary research areas or fields such as business psychology, neuroscience, and behavioral economics. There is a great potential for development of research in such areas and also areas that have so far not been identified.

George Loewenstein, Kathleen Musante, and Joshua A. Tucker identify four categories of challenges faced by researchers who pursue interdisciplinary social science research. First, culture challenges which include the difference in methodological approaches and lexicons used across different disciplines. Second, professional challenges such as hurdles to recognition, publication, and promotion in interdisciplinary fields. Larger number of co-authors is generally looked down upon by traditional disciplinary journals. Third, training challenges. Disciplinary models and administrative skills dominate existing scholarly training and long-term research projects. Last, financial challenges or resource constraints. The present funding pipelines largely support disciplinary research. Interdisciplinary research generally requires significantly larger grants for its sustenance. However, interdisciplinary research is generally awarded much lesser funds than disciplinary research projects. Particularly when establishing large-scale social science research labs, which currently almost do not exist. Interdisciplinary research could substantially benefit from such initiatives. They further identify potential solutions [12]:

- Establishing high-profile funding opportunities and explicit interdisciplinary journals including specialized research labs for social sciences.
- Creating novel models of graduate and post-graduate training facilities and programs that bridge disciplinary divides.
- Implementing novel approaches to train interdisciplinary scholars to review related grants and research proposals.
- Establishing intellectual spaces such as summer schools where workshops that aim to address common problems faced by different disciplines. This will encourage collaboration and mutual assisted learning through co-assistance. It will further lead to targeted specific research collaborations.
- Identifying and establishing novel models that structure research funding in various social science disciplines and schools which are more conducive to interdisciplinary social science research.

Social science and humanities are essential fields of research. Irrespective of the arguments and contentions against the fields, the relevance of these fields endures. Below listed are reasons as to why social science will not lose relevance in the universities of the future [17].

Social science researchers and scientists put forth alternate futures that humankind could pursue. It can open up discussions and debate that could shape the collective future. The present range of ethical, social, and legal issues faced by societies require refined social science research. Social science also helps us make sense of our finances. For instance, economic crisis can be better explained by not just economists but also sociologists, psychologists, and political scientists. Social scientists contribute to physical and psychological well-being. From medical statisticians to public health experts to sports sociologists, social science practitioners are an essential part of healthcare. Social science can improve future and living standards. It can change the world for the better. Debates on peace, feminism, ecology, social movements, and so on offer new perspectives and ways of understanding. This will broaden one's horizons and helps understand how the world functions.

Andrew Abbott puts forth four individual transitions that are possible in the fate of social sciences and humanities: first, the emergence of a scientific model of knowledge along the "one size fits all" line; second, the increasing neoliberal management in the academic world; third, the shift from print to image or digital (from a "complex argument to a simplified assertion" or from "discursive symbols to presentational ones"). The fourth transition, that is mentioned, is the growing disparity between sophistication of empirical social science and simplicity of its normative reasoning. The first two scenarios are in Braudelian terms; the third is in Braudelian structure and the fourth transition is a conjunctural one [1].

5 The Future: Technology, Interdisciplinarity, and Dynamism

The social sciences and humanities require an increased investment in normative theory, wherein the latter is predominantly process-based in order to get the researchers beyond the simplicities of the present normative ontology. The current normative ontology is not adept in tackling the complexities of the contemporary societies and communities. As Abott argues, a normative social ontology enables the conceptualization of normatively governed social processes which can appreciate, understand, and govern the constant dynamic nature of the society and populations.

Since the twentieth century, technological development and scientific advancement have caused remarkable capabilities which significantly influence social lives. Energy, telecommunications, electronics, transportation, medicine, and defense, among many other advancements, have altered the way humans live and function. Coupled with the COVID-19 pandemic and the resultant remote-working practices, significant alterations have occurred in the social contexts. A hybrid learning model features as a crucial part in any educational institution's future plans. This calls for new trajectories of fundamental social science research. Interdisciplinary and multidisciplinary research programs will be able to throw light on various developments and global challenges. Furthermore, it will help in the context of global challenges and enable practitioners to seize potential opportunities.

Historically, universities were expected to have just one duty – to provide knowledge and skills to the students. In this day and age, do we still expect universities to pump out individuals who are prepared for the job market? One can argue that this expectation is old-fashioned. Universities must also evolve and adapt to the changing social scenarios. It must be able to incorporate the demands and desires of the job markets. This does not mean that subjects that may not be pre-professional or pure science must supersede social sciences and humanities.

In the Future University, it could no longer be the university's responsibility to guarantee jobs for the students. Future University must move beyond the primitive debate linking job opportunities with social sciences and humanities. In fact, the stronger together one clubs job opportunities and social sciences, the more one humiliates social sciences and humanities. Future University will also create individuals with more digital and technological literacy. These individuals will be prepared to work but will also be adept in contributing to different fields. The constant questioning of the purpose and benefit of universities will be a catalyst to the changes that will culminate in the university of the future. It will push the education sector on to a trajectory to an ecological model. An ecological university is one that is deeply integrated into the society (communities and universities) around it, ensures easy availability of its knowledge resources, and engages proactively to manifest a better world.

While there are many jobs that match existing job profiles and opportunities, various interdisciplinary programs and traditional social science courses face the issue of a lack of employment opportunities. While some do undeniably contribute to national development, it is tomfoolery to expect the universities of the future to remain on their predecessors' paths. Undoubtedly, Future Universities will explore the realms beyond the primitive debate that simply links a university degree with a job opportunity. Moreover, the main outcome of Future University programs will no longer just be knowledgeable students.

The way ahead for social sciences would largely remain strengthening existing and creating new interdisciplinary programs. Its advantages – catering to students with a wide range of professional and educational backgrounds, providing multiple options for branching out and specialization, and offering a wide range of career options – will maintain the desirability of interdisciplinary programs [29].

Social sciences and humanities have been undermined in today's university. The rollbacks in funding and closing down of entire departments lend themselves evidence to this argument. The false perception that pure science is the only solution did harm social sciences and its related fields. Nevertheless, as observed in the paper, the future of social sciences and humanities need not be glum. Utilizing interdisciplinary and multi-disciplinary approach to bridge the gap between social science and universities has worked. The interdisciplinary approach did help social sciences and humanities survive. In the Future University, the approach will need a revamp and strengthening. Nevertheless, global problems require and re-emphasize the importance of social sciences. Human insights and social data are crucial to address the impending issues.

References

1. Abbott, A. (n.d.). *The future of the social sciences*. University of Chicago. https://home.uchicago.edu/aabbott/Papers/Marc%20Bloch%20Lecture%20Pre%20Trans.pdf. Last accessed January 2022.
2. Alexander, B. (2020). *Academia next: The futures of higher education*. John Hopkins University Press.
3. Bates, R. H. (1997). *Area studies and the discipline: A useful controversy?* American Political Science Association.
4. Belkin, D. (2020, April 10). College students demand coronavirus refunds. *The Wall Street Journal*. https://www.wsj.com/articles/college-students-demand-refunds-after-coronavirus-forces-classes-online-11586512803. Last accessed January 2022.
5. Bilgin, P. (2004). Is the 'orientalist' past the future of Middle East studies? *Third World Quarterly, 25*, 423–433.
6. Blakemore, E. (2015, October 5). *Japanese universities are shuttering social sciences and humanities departments*. https://www.smithsonianmag.com/smart-news/japans-government-asks-all-universities-shutter-humanities-and-social-sciences-departments-180956803/. Last accessed January 2022.
7. CEPD. (2012). In Centre for Education Policy Development (CEPD). Centre for Education Policy Development (CEPD). Republic of South Africa: Higher Education and Training Republic of South Africa (Ed.), *The future of humanities and social sciences in South African universities conference*.

8. Dean, A. (2015, September 26). *Japan's humanities chop sends shivers down academic spines.* The Guardian. https://www.theguardian.com/higher-education-network/2015/sep/25/japans-humanities-chop-sends-shivers-down-academic-spines. Last accessed January 2022.
9. EL-Tohamy, A. *Covid-19 poses new challenges for social scientists.* Survey Finds. https://www.al-fanarmedia.org/2021/07/covid-19-poses-new-challenges-for-social-scientists-survey-finds/. Last accessed January 2022.
10. Kiley, K. *Another liberal arts critic.* https://www.insidehighered.com/news/2013/01/30/north-carolina-governor-joins-chorus-republicans-critical-liberal-arts. Last accessed January 2021.
11. Lochen, Y. (2012). Commitment and imagination in the social sciences: A concern for the future state of sociology. https://www.tandfonline.com/doi/abs/10.1080/13511610.1994.9968420?journalCode=ciej20. Last accessed November 2022.
12. Loewenstein, G., Musante, K., & Tucker, J. A. (2019). *Future directions in social science a workshop on the emergence of problem-based interdisciplinarity* (Workshop). Virginia Tech Applied Research Corporation.
13. Manski, C. F. (1993). Identification problems in the social sciences. *Sociological Methodology, 23,* 1–56.
14. Mazur, E., & Kerrey, B. Higher Ed's coronavirus opportunity. *The Wall Street Journal.* https://www.wsj.com/articles/higher-eds-coronavirus-opportunity-11589139956. Last accessed January 2022.
15. Mirssepassi, A., Basu, A., & Weaver, F. (2003). *Localising knowledge in a globalizing world.* Syracuse University Press.
16. Murphy, J. *Nearly four dozen faculty jobs to be cut; tough times in the Pa.* State System of Higher Education. https://www.pennlive.com/midstate/2013/11/nearly_four_dozen_state_univer.html. Last accessed January 2021.
17. Palgrave Macmillan. (2015). *10 reasons why we need social science.* https://www.palgrave.com/gp/campaigns/social-science-matters/10-reasons-for-social-science. Last accessed January 2022.
18. Schwab, K. (2016). *The fourth industrial revolution.* Crown Business.
19. *The Fourth Industrial Revolution: What it means, how to respond.* World Economic Forum. https://www.weforum.org/agenda/2016/01/the-fourth-industrial-revolution-what-it-means-and-how-to-respond/. Last accessed March 2022.
20. Shah, H. *Global problems need social science.* https://www.nature.com/articles/d41586-020-00064-x. Last accessed 16 Jan 2022.
21. Shannon, V. P. (2007). American orientalism: The United States and the Middle East since 1945 (review). *Journal of Cold War Studies, 9*(4), 149–151.
22. Silva, K. T. (2020). Opportunities and challenges for social sciences in the aftermath of the COVID-19 pandemic. *Sri Lanka Journal of Social Sciences, 43,* 1.
23. The Wall Street Journal. *Coronavirus prompts colleges to send students home.* https://www.wsj.com/articles/coronavirus-prompts-colleges-to-send-students-home-11583862936. Last accessed January 2022.
24. *Coronavirus pushes colleges to the breaking point, forcing 'hard choices' about education.* The Wall Street Journal. https://www.wsj.com/articles/coronavirus-pushes-colleges-to-the-breaking-point-forcing-hard-choices-about-education-11588256157. Last accessed January 2022.
25. Tworek, H. (2013, 18 December). *The real reason the humanities are 'in crisis'.* https://www.theatlantic.com/education/archive/2013/12/the-real-reason-the-humanities-are-in-crisis/282441/. Last accessed January 2022.
26. Wachbrit, G. (2019, November 26). *Identifying the challenges of social science's newest technology.* https://www.socialsciencespace.com/2019/11/identifying-the-challenges-of-social-sciences-newest-technology/. Last accessed January 2022.
27. Wood, E. J., et al. (2020). *Resuming field research in pandemic times.* Social Science Research Council.

28. World Economic Forum. *World Economic Forum Jobs Reset Summit: Building back broader in the economic recovery.* https://www.weforum.org/press/2021/06/world-economic-forum-jobs-reset-summit-building-back-broader-in-the-economic-recovery-1997e346c4/#:~:text=The%20Jobs%20Reset%20Summit%20and%20associated%20initiatives%20are%20dedicated%20to,growth%20models%2C%20taxati. Last accessed March 2022.
29. Zweiri, M., & Al-Qawasmi, F. (2019). *Advanced interdisciplinary programs place new demands on professors.* Al Fanar Media. https://www.al-fanarmedia.org/2019/04/advanced-interdisciplinary-programs-place-new-demands-on-professors/. Last accessed January 2022.

Impact of the Industry 4.0 on Higher Education

Thafar Almela

1 Introduction

The term fourth industrial revolution (IR 4.0) was originally invented and introduced during the Hannover Trade Fair in Germany in 2011 by three engineers: Henning Kagermann, Wolfgang Wahlster, and Wolf-Dieter Lukas [1]. Kagermann et al. described the concept of Industry 4.0 as the technical integration of cyber-physical systems (CPS) into manufacturing and Services. These CPS comprise smart machines, storage systems, and production facilities capable of autonomously exchanging information, triggering actions, and controlling each other independently. This will have implications for value creation, business models, downstream services, and work organization [2]. In 2016, the World Economic Forum (WEF) referred to IR 4.0 as the convergence of technologies so that the line between the physical and digital spheres becomes indistinct. Artificial intelligence (AI) and robotics, the Internet of things, additive manufacturing, neurotechnologies, biotechnologies, and virtual and augmented reality are identified as some of the key technological drivers of IR 4.0 [3, 4].

Worldwide, many countries are preparing for entering the new era, and higher education (HE) has a leading role in shaping the transitions necessary to adjust to the automation economy carried by IR 4.0. While the shifting toward digitization holds great promise, it also poses major challenges in the reformation of HE that requires strategic planning to comply with the upcoming changes. A study report, "Towards a 2030 Vision on the Future of Universities in Europe," prepared by the European Commission's Directorate-General for Research and Innovation set out the vision of the transformation of universities in Europe as well as the pathway

T. Almela (✉)
Department of Oral and Maxillofacial Surgery, College of Dentistry, University of Mosul, Mosul, Iraq
e-mail: tkdalmela@uomosul.edu.iq

toward the achievement. The vision statement made in this report stated that by 2030, Europe's universities will be world leading in research and innovation, grounded in disciplinary excellence, and interdisciplinary approaches to address complex problems as well as retain a high degree of autonomy and excellence in research and innovation [5].

Such transformation imposes identifying a set of transferable skills that will prepare all students, regardless of their discipline or professional field, for working in the future. These skills often involve competencies related to innovation, problem solving, digital literacy, and lifelong learning. The need for such skills is inescapable in confronting major challenges facing the world that cannot be addressed by the traditional education system designed to meet the needs of past industrial revolutions. Perhaps, the next era is more about *what* students can do with their knowledge after leaving the university than *how much* knowledge has they obtained.

While Industry 4.0 is still evolving, professionals involved in education must embrace this transformation through high-quality learning—"Education 4.0." Miranda et al. defined Education 4.0 as the implementation of new learning methods, novel pedagogic and managerial tools, as well as smart and sustainable infrastructure in HE institutions, which is primarily supplemented by new and developing technologies. Combining these resources will enable the students to obtain the desirable critical competencies needed for future work [6]. However, this trend in education requires active collaboration among HE institutions, industry, governmental agencies, and student/alumni organizations to stay updated on the emerging competence needs.

Traditional education has contributed greatly to the current levels of the industrial evolution and technological advancement. However, for HE to deliver future generations with the right set of skills and knowledge; an imperative question has to be asked regarding how the education system can be reformed for effective students' preparation where technology is swiftly changing and becomes entrenched within society. This chapter discusses the significant impact of industry 4.0 on HE, illustrates some successful examples and experiences from countries that have already embraced education 4.0, and explains the challenges arising from education 4.0.

2 Limitation of the Current HE System

HE is considered one of the key players in the development of national and international economies. University-industry connections can result in the growth of a high-quality workforce, increasing innovation, upskilling and reskilling of existing workforces, and minimizing unemployability [7]. The critical role of HE in shaping societies raises questions about the adequacy of current education systems in keeping pace with the changes brought by the advent of IR 4.0. In general, the limitations of the existed HE can be summarized under two main themes: quality and accessibility.

2.1 Quality of HE

The pedagogical approach is the essence of the learning process. Instructor-centered education is one of the main drawbacks of the traditional HE systems in many developed and developing countries. Instead of interactive approaches that promote the critical and individual thinking needed in today's innovation-driven economy, the HE largely depends on passive modes of learning, direct instruction, and memorization. The ineffectiveness of teaching through the transmission of knowledge has been confirmed through years of pedagogical research [8]. However, the current educational system still sets the delivery of knowledge as the goal of learning and emphasizes the amount of knowledge transmitted from teachers to students who will be evaluated through exams based on the mere repetition of memorized information. Student-centered learning (SCL), by contrast, which includes experiential learning, participatory learning, and problem-solving approaches, ensures that the targeted learning outcomes of a course component are attained in a way that encourages critical thinking and transferable skills. The key concept underlying SCL is promoting an autonomous personalized education because HE is diverse, that is, different students have a different educational background, different interests and skills, and different learning style [9]. Accordingly, the "One-Size-Fits-All" mode of learning seems irrelevant in terms of enabling the students rather than telling them much information that may become obsolete even before it can be used.

2.2 Accessibility to HE

While there are several barriers to HE such as geographical mobility, discrimination, crisis, and emergency, the financial burden including the high tuition fees and learning expenses may constitute a major drawback of the conventional HE. In many countries, learning is restricted to university buildings and it is affordable to those who can pay the cost of learning in highly ranked institutions. A report created by the UNESCO International Institute for Higher Education in Latin America and the Caribbean (UNESCO IESALC) investigated the period from 2000 to 2018 to map and analyze the main trends and global access to HE [10]. The report clearly showed the disparities and the significant differences in access to HE, particularly for income groups. The poorest population continues to lag behind, with 10% access to HE in 2018 compared to 77% of the higher income sector in the same year. The upper middle-income showed the highest increase in participation rates while the lowest has been in low-income countries. Between 2000 and 2018, the gross enrolment ratio (GER) in upper middle-income countries increased by more than 200%. There seems to be a strong relationship between gross domestic product (GDP) per capita and tertiary GER (see Fig. 1). However, it must be noted that despite the universal access to HE has been increasing worldwide over the past two decades and the global HE gross enrollment ratio increased from 19% to 38%, there is a gap

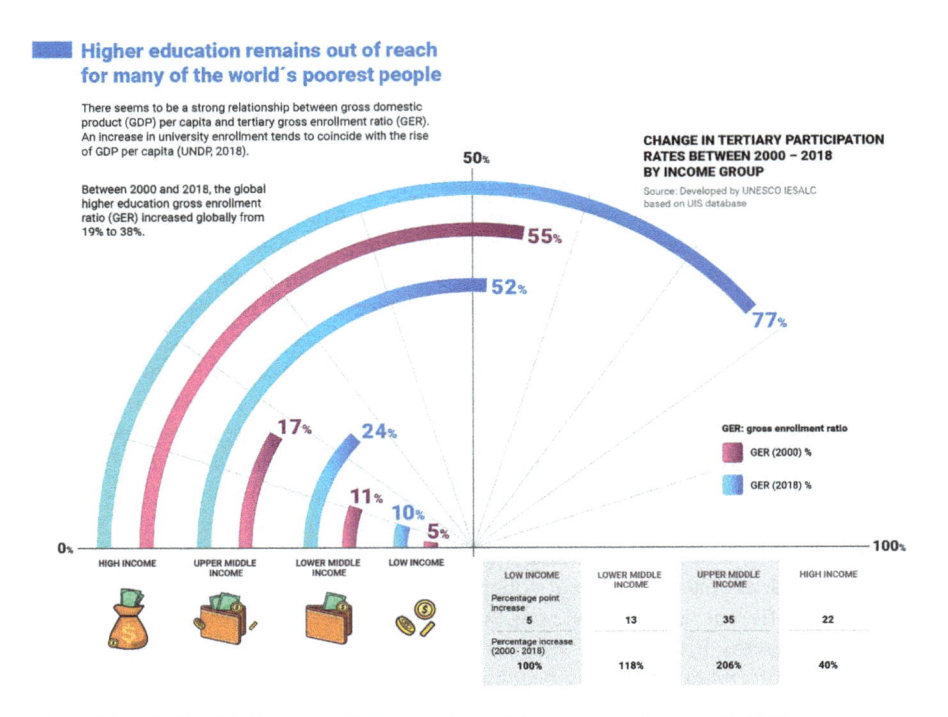

Fig. 1 The relationship between GDP per capita and the gross enrolment ratio [10]

between enrollment and graduation rates. For example, in the USA, more than one million students drop out of college every year. More than two-thirds of college dropouts are low-income students, with family-adjusted gross income (AGI) under $50,000. High-income students with a family AGI of $100,000+ are 50% more likely to graduate than low-income students [11]. Such findings reveal that not all segments of society are equally able to benefit from the current education system.

As we embark on IR 4.0, there is a high ambition that technologies will play an important role in addressing the limitations of the existing HE. On the other hand, the automation and digitalization carried out by Industry 4.0 will result in marked economic changes such as the elimination of lower-skilled jobs [12]. As such, to cope with the remarkable change in the industry, HE should be transformed to provide broader workforce skillsets and job-specific capabilities, closing the IT skills gap, and offering new formats for continuing education [13]. According to the WEF estimate, as much as US$11.5 trillion could be added to global GDP by 2028 if countries succeed in better preparing learners for the needs of the future economy [14]. Yet, this requires a significant change in the education system, institutional structures, and strategic leadership. Moreover, such shifting in HE cannot be achieved in isolation from the education system in primary and secondary schools. Based on the framework published by the WEF report in 2020 "Schools of the Future: Defining New Models of Education for the Fourth Industrial Revolution," eight critical characteristics in learning and experiences for schools have been identified to define the Education 4.0 model that meets the needs of the future (see Fig. 2) [15].

3 The Impact of Previous Industrial Revolutions on Higher Education

Every industrial revolution sets an important pattern for education to cover the unmet demands. As such, HE has evolved from Education 1.0 to the current educational paradigm of Education 4.0.

The First Industrial Revolution (1760–1840) involved the transition from hand production to mechanical production powered by steam. The HE in this period demanded a new kind of curriculum and more diverse degree options. The widespread adoption of the German university model of research universities allowed the growth of post-graduate research in the US and worldwide. The president of Harvard, Charles W. Eliot, described this type of education as "The New Education" [16].

The Second Industrial Revolution (1860–1900) launched what was described as a "new economy" [17]. It is associated with new manufacturing technologies based on electricity. In the USA and Europe, this period witnessed increased access to HE and remarkable expansion of a variety of HE institutions funded by both public and private sources. This resulted in a surging of discoveries carried by powerful technology. The goal of these institutions was to produce a steady stream of newly well-trained workers prepared for the "practical avocations of life" such as agriculture and the mechanic arts [18].

The Third Industrial Revolution (1980s–1990s) is generally attributed to computerization and web-based interconnectivity. Within the Third Industrial Revolution, there was a large increase in the diversity of campuses and the

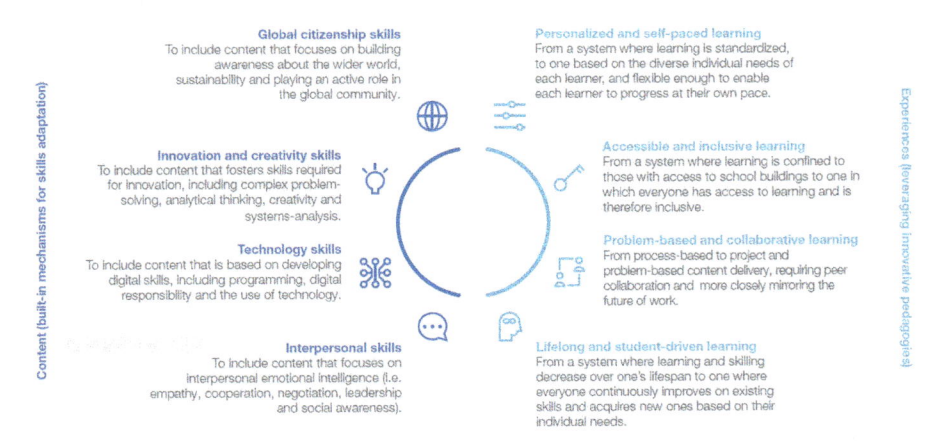

Global citizenship skills
To include content that focuses on building awareness about the wider world, sustainability and playing an active role in the global community.

Innovation and creativity skills
To include content that fosters skills required for innovation, including complex problem-solving, analytical thinking, creativity and systems-analysis.

Technology skills
To include content that is based on developing digital skills, including programming, digital responsibility and the use of technology.

Interpersonal skills
To include content that focuses on interpersonal emotional intelligence (i.e. empathy, cooperation, negotiation, leadership and social awareness).

Personalized and self-paced learning
From a system where learning is standardized, to one based on the diverse individual needs of each learner, and flexible enough to enable each learner to progress at their own pace.

Accessible and inclusive learning
From a system where learning is confined to those with access to school buildings to one in which everyone has access to learning and is therefore inclusive.

Problem-based and collaborative learning
From process-based to project and problem-based content delivery, requiring peer collaboration and more closely mirroring the future of work.

Lifelong and student-driven learning
From a system where learning and skilling decrease over one's lifespan to one where everyone continuously improves on existing skills and acquires new ones based on their individual needs.

Content (built-in mechanisms for skills adaptation)

Experiences (leveraging innovative pedagogies)

Fig. 2 The WEF Education 4.0 framework [15]

globalization of academic research accelerated by online technologies. In addition, access to HE rose prominently with an intensified commitment to large-scale HE across the world. For example, in 2000, approximately 25.6% of the US population aged 25 and above had graduated from HE institutions which significantly increased from 1960 when the US graduated students were only 7.7% (see Fig. 3) [19].

Probably, the move toward online education is one of the remarkable achievements of the third Industrial Revolution that continue to develop in IR4.0. Massive Open Online Courses "MOOC" were introduced as a new approach to making university education affordable for millions of students who have previously deprived form education. The development of online education is likely to build the students' knowledge and skills asynchronously by combining high-quality synchronous, in-person learning environments with online technologies (Gleason 2018). The market size of online education has significantly increased by the growing application of cloud-based solutions and massive investments by major companies to enhance the security and reliability of cloud-based education platforms. According to the research and market report, the global online education market reached US$187.877 billion in 2019 and it is projected to increase up to US$319.167 billion in 2025 [20]. Increasing use of the Internet in many regions across the globe is a major factor driving the market growth. This has prompted many educational technology companies to partner with world-leading universities and institutions to deliver online education. For example, 2U, Inc., an American educational technology company founded in 2008 as an online program manager supplying its client institutions with cloud-based software, coursework design, and infrastructure support. The company is contracting with major colleges and universities to build, deliver, and support online degree and non-degree programs. In 2021, 2U announced its acquisition of

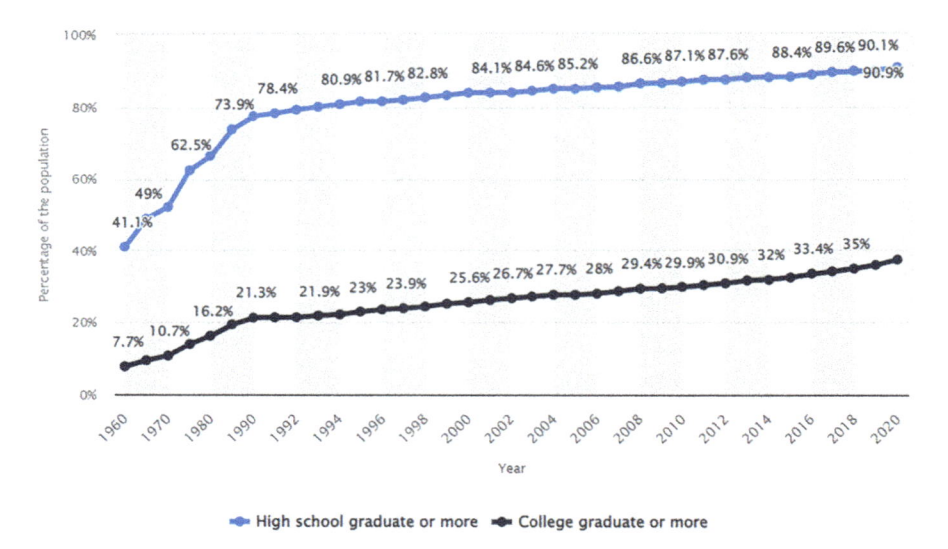

Fig. 3 Educational attainment distribution in the United States from 1960 to 2020 [19]

the edX, a MOOC provider created by MIT and Harvard in 2012. 2U was named by Forbes as one of the ten "Start-Ups Changing the World" [21].

The exact impact of IR 4.0 technologies on HE is still unknown; yet, they will likely bring profound change. However, it has been noted by economists that usually there is a significant lag time between the introduction of new technology and the full embracement of that technology when it becomes possible to obtain measurable impacts on productivity. This lag between training and experimentation with new technology to widely disseminate throughout society is called a "productivity paradox" [17]. Historically, the results suggest that time is required to fully realize changes within society and the impacts of technology on education. Perhaps, this rule is also applicable for IR4.0 and the major changes in education arising from this revolution may span several decades until it spread globally.

4 Requirements of Education 4.0

4.1 Curriculum

Curriculum revision needs to enable the students to envisage the importance of what they are learning with aspects of their lives and the real world. This entails *relevant* content material that connects the learned knowledge and skills with real-world problems and applications [22]. Stanford University can be considered a pioneer in this context by developing an innovative curriculum that engages the students in real problems. For example, the life science curriculum is reshaped to allow students enrolled in the new engineering biology course to solve real problems in medicine, public health, and environmental management through digital designing and bioprinting of the life forms [23]. Similar innovations within chemistry include courses and degree programs in Green Chemistry, which blends chemistry, biology, and environmental science to allow students to engage in practical environmental problems such as synthetic fuels, bioplastics, toxicology, and techniques to reduce pollution [24].

Another essential requirement of the education 4.0 curriculum is to be *Interdisciplinary*. Obtaining universal skills, working in multidisciplinary teams, completing interdisciplinary projects, and providing interdisciplinary thinking are prerequisites of Industry 4.0. Therefore, the single discipline nature of traditional education seems irrelevant to the nature of IR4.0. and an urgent modification of the current program structures and/or course syllabuses is essential for future education [25]. In recent years, new institutions have been created with more global and interdisciplinary curricula. One example is Yale-NUS College in Singapore, developed by Yale University and the National University of Singapore to provide a residential liberal arts college within Asia. Yale-NUS College offers an interdisciplinary curriculum that features literature and philosophy from both Eastern and Western cultures, a range of interdisciplinary science courses, and quantitative reasoning [26].

Crucially, the reformation of the curriculum should be accompanied by a comparable transformation of the assessment because it encompasses the end product of the educational process. Technology provides opportunities to test knowledge and skills in a more realistic and motivating way than handwritten examinations, which can appear irrelevant outside the academia. In 2019, the Joint Information Systems Committee (JISC), the UK's expert body for digital technology and digital resources in higher education, further education and research, convened a meeting for a wide range of experts and contributors to explore the future of assessment in universities and colleges. The resulted report indicated that emerging technologies can develop smarter, faster, and fairer assessments as well as address issues such as assessing the right things, assessing in right points in the learning process, involving a sustainable workload, and being trusted. In addition, the report set five principles and targets for 2025 to achieve digital transformation of the assessment by making it more *Authentic, Accessible, Appropriately automated, Continuous,* and *Secure* [27]. An example of applying this new concept of assessment is MySkills (https://myskills.org.uk/) project led by Glasgow College and involving the Scottish Qualifications Authority. This digital qualification project is investigating the use of blockchain to support micro-credentialling and allow employers to verify qualifications. The project started in October 2018 and has produced a model for how using blockchain can achieve the transition to digital certificates in education.

4.2 Skills

The skills required for the automation economy are not only different from those that have been emphasized by HE institutions in the past, but they are changing rapidly. In terms of the overall scale of demand for various skills in 2020, the WEF survey in 2016 indicated that less than 1 in 20 jobs (4%) will have a core requirement for physical abilities compared to more than one third (36%) of all jobs across all industries are expected to require complex problem-solving as one of their core skills (see Fig. 4) [28]. However, the WEF report published in 2020 revealed that the skills continuing to grow in prominence by 2025 include analytical thinking and innovation. Although problem-solving has remained at the top of the agenda, it becomes apparent that the demand is growing for other newly emerging skills in self-management such as active learning, resilience, stress tolerance, and flexibility (see Fig. 5) [14].

Another report created by the Universities of the Future (UoF), (https://universitiesofthefuture.eu/), a project cofounded by Erasmus program of the European Union (EU) to address the educational needs arising from Industry 4.0 in Europe, identified the skills required for succeeding in the Industry 4.0 environment and figured the framework for working life skills into *discipline-specific competencies* and *transferable skills* (see Fig. 6) [29].

As needed skills are rapidly evolving, the HE sector has to adapt. The digital transitions require research and innovation in close cooperation with the related

Impact of the Industry 4.0 on Higher Education 157

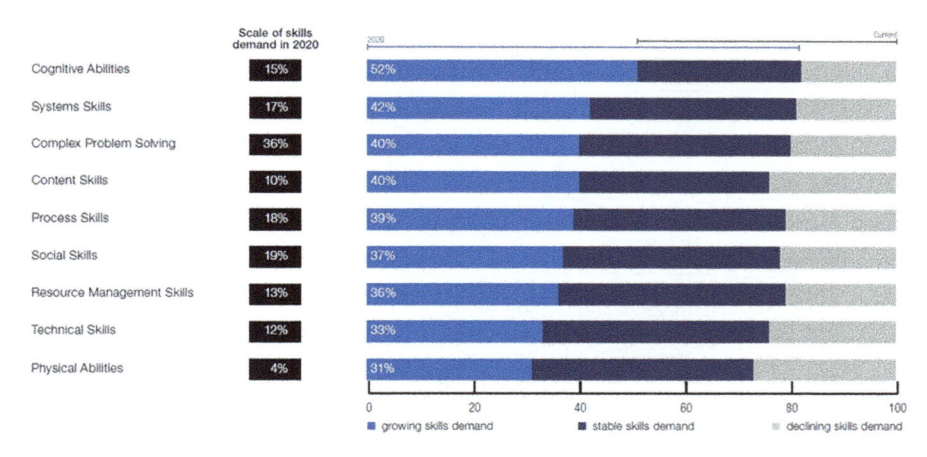

Fig. 4 Change in demand for core work-related skills, 2015-2020, all industries [28]

Scale of skills demand in 2020:
- Cognitive Abilities: 15%
- Systems Skills: 17%
- Complex Problem Solving: 36%
- Content Skills: 10%
- Process Skills: 18%
- Social Skills: 19%
- Resource Management Skills: 13%
- Technical Skills: 12%
- Physical Abilities: 4%

2020 vs Current (growing / stable / declining skills demand):
- Cognitive Abilities: 52%
- Systems Skills: 42%
- Complex Problem Solving: 40%
- Content Skills: 40%
- Process Skills: 39%
- Social Skills: 37%
- Resource Management Skills: 36%
- Technical Skills: 33%
- Physical Abilities: 31%

A. Relative importance of different skill groups

(Decreasing / Stable / Increasing — Share of companies surveyed (%))
- Critical thinking and analysis
- Problem-solving
- Self-management
- Working with people
- Management and communication of activities
- Technology use and development
- Core literacies
- Physical abilities

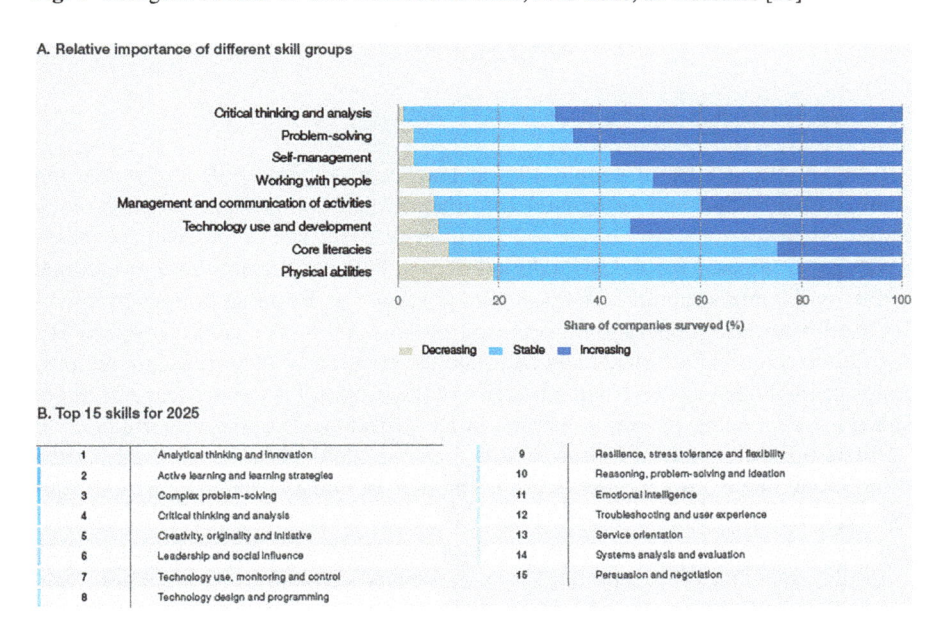

B. Top 15 skills for 2025

#	Skill	#	Skill
1	Analytical thinking and innovation	9	Resilience, stress tolerance and flexibility
2	Active learning and learning strategies	10	Reasoning, problem-solving and ideation
3	Complex problem-solving	11	Emotional intelligence
4	Critical thinking and analysis	12	Troubleshooting and user experience
5	Creativity, originality and initiative	13	Service orientation
6	Leadership and social influence	14	Systems analysis and evaluation
7	Technology use, monitoring and control	15	Persuasion and negotiation
8	Technology design and programming		

Fig. 5 Perceived skills and skills groups with growing demand by 2025 [14]

industries and stakeholders. For this reason, many countries have already set a strategic plan for the advancement of digital education. For example, the EU launched two initiatives that go in this direction: a European Strategy for Universities and a Commission proposal for building bridges for effective European higher education cooperation. The first initiative involves the forward-looking strategic vision called "Digital decade" for the development of the digital economy and the transformation of European businesses by 2030. The action plan of Digital Decade sets ambitious

```
                        ┌─────────────────────┐
                        │  Working life skills │
                        └─────────────────────┘
```

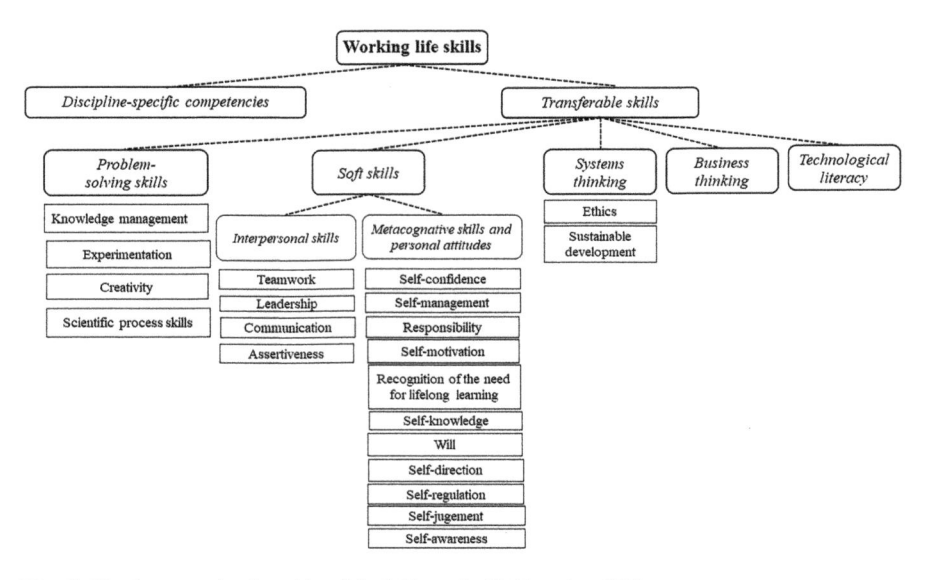

Fig. 6 The framework of working life skills set by UoF project [29]

targets, aiming at 80% of people with at least basic digital skills and increasing number of qualified and competent Information and communication technology employed professionals in Europe to 20 million by 2030. The Commission proposes a "Digital Compass" for the EU's digital decade that evolves around four cardinal points: skills, infrastructures, business, and government. Building bridges for effective European HE cooperation, on the other hand, aims to enable European HE institutions to cooperate closer and facilitate the creation of joint transnational programs or joint degrees. The main objective of the proposal is to encourage Member States to support the provision of high-quality life-long learning opportunities to facilitate upskilling and reskilling, with a focus on the most in-demand areas such as artificial intelligence, cybersecurity, Internet of things, and augmented reality [30].

In Asia, Indonesia is one of the countries that adopt an innovative pedagogy to foster the student's skills required for an automated economy. A comprehensively evaluated development project "Modernizing Indonesian Higher Education with Tested European Pedagogical Practices INDOPED" was European Commission-funded project in 2015–2019 (www.indoped.eu). In INDOPED, 5 European Universities provided mentoring support to 5 Indonesian universities to test student-centered active learning methods. Over 100 Indonesian lecturers engaged thousands HE students in pedagogical pilots in a wide range of faculties, e.g., business, languages and arts, agriculture, and medicine. The results of the INDOPED project are motivating. The objective external evaluators state that "INDOPED project has brought student-centered innovation-oriented teaching methods to Indonesia. The project does not only provide the teachers with new pedagogy skills but also allows the university teachers and administrators to reflect upon their teaching and see the relations between the goal they try to achieve, i.e., preparing their students to enter the workforce in Industrial Revolution 4.0 era, and the strategies they apply to achieve the goal" [31].

Vietnam is another southeast Asian country that sets a remarkable example in utilizing an innovative pedogeological strategy to foster students' new skills. Turku University of Applied Sciences (TUAS), one of the biggest universities in Finland, is coordinating joint efforts of 6 Vietnamese Universities and 5 European Universities in developing the modern "Smart Sustainable Vietnamese Cities SSVC" study module. The SSVC module is implemented in a blended learning by including online, onsite, and project learning enriched with stakeholder cooperation (www.saunac.eu). This European Commission-funded operation enabled the participating Vietnamese universities to utilize the power of students in creating many inspiring ideas such as proposing improvement of urban flooding management tools.

These pedagogical novelties should act as signposts for all education providers. Adopting new pedagogical strategies leads to building necessary skills for the future workforce as well as creating new kinds of cooperation (e.g., between lecturers and students; between universities and companies, public organizations, and non-governmental organizations) which increases trust and paves the way for the new era in education.

4.3 Innovation and Entrepreneurship

Disruptive technologies and processes are invented when revolutionary innovations dominate. Within this context, the challenge becomes how to teach the students so that they become innovators. According to Schwab [4], innovation and entrepreneurship can be classified under two research areas:

- Innovative materials and tools that enable the developing new products such as nano-technological materials.
- Innovation/entrepreneurship business models suggest that companies should use external ideas when they look for technological advances.

Countries worldwide realized that innovation and entrepreneurship are the main sources of job growth. In 1999, the 3rd National Conference on Education was held in China aiming for promoting quality-oriented education and innovative capacity. Since then, an increasing percentage of the national budget has been allocated to education in response to the need to create a culture of innovation and entrepreneurship has been significant. In a few short years, Chinese Innovation yielded many leading companies in different fields such as artificial intelligence and machine learning. However, the main challenge is continuance which involves creating a culture or ecosystem of creativity, critical thought, and entrepreneurial mindset. These environments are created by practices developed early in life and deeply influenced by the social, economic, and educational institutions [32]. As nurturing innovative capacity requires long-term and sustained efforts, China's strategy for the development of corporate innovation undergoes certain transformations started in 2009 with annual financing of $6.5 billion and aimed at stimulating innovative entrepreneurship by increasing the innovative activity of small- and medium-sized enterprises (SME). In addition to this program, the government launched the SBIR (The Small Business Innovation Development and Research) program in 2010. The aim was to attract private businesses to solve specific scientific and technical problems faced by 10 largest federal ministries and national agencies. By 2016, the Chinese government spent more than $56 billion in total to stimulate mass innovation entrepreneurship [33]. Moreover, the Ministry of Education (MOE) has issued specific policies to support innovation and entrepreneurship by allowing students to take one year off to pursue an entrepreneurial venture [34]. In 2015, the MOE suggested all universities provide elective courses in entrepreneurship for credit (compulsory and selective) to all students [35].

Singapore is another example of countries that has already framed and implemented its national policy regarding Education 4.0 innovation. In 2016, the Committee on the Future Economy was convened to develop economic strategies for the next decade. The report outlines Singapore's economic strategies for IR 4.0 preparation and lays out an innovative 7 Strategy Plan. The three strategies relative to HE are as follows: deepen and diversify the international connections, acquire and utilize deep skills, and build strong digital capabilities [36]. The first education strategy, for instance, is pursued through the Global Innovation Alliance initiative to form alliances between Singapore's institutes of higher learning and major innovation hubs around the world. As a result of this strategy, the National University of Singapore's (NUS) Overseas Colleges started internship programs in Beijing, Israel, Lausanne, Munich, New York, Shanghai, Silicon Valley, Singapore, and Stockholm. With the holistic education systems that teach students how to learn, rather than what to learn, Singapore serves as a model to nations around the world wanting to prepare their workforce with ways of thinking and working that are in demand in IR 4.0.

Intrapreneurship is a model for promoting innovation in companies where management supports entrepreneurial thinking within the company. This model of innovation can be carried out through collaboration with the universities where the researchers work in companies both to learn the current practices and to give industry insights into the latest research. For example, Post Docs in Companies "podoco.fi" is a joint initiative of universities, industry, and foundations started in Finland in 2015. The PoDoCo program aims to promote academic research, supporting long-term competitiveness, and the employment of young doctors in Finnish companies. PoDoCo program is aimed for all branches of industries and all disciplines, e.g., natural sciences, engineering and technology, medical and health sciences, agricultural sciences, social sciences, and humanities. After the research period, the company employs the postdoc to deepen the research results and to create company-specific insight. This results in a mutual benefit between industry and academia where academic research is supporting the long-term competitiveness and strategic renewal of Finnish companies and the scholars are obtaining the industrial experience.

5 Challenges of Education 4.0

5.1 Lack of Opportunity and Passion

While there is a high excitement for change in education, the pace of change is still slow. This is not surprising due to the challenges in this area. The readiness of organizational culture is of utmost importance for educational transformation. This is inevitably associated with resources, funding, awareness, staff time, and willingness to engage. The statistics from 61 HE and further education (FE) organizations in the UK showed that only 34% of HE teaching staff and 36% of FE teaching staff were offered regular opportunities to develop their digital skills. In addition, just 33% of FE teaching staff and 27% in HE agreed that they receive guidance about the digital skills they are expected to have as a teacher, and only 17% of FE and 15% of HE teaching staff are using assistive technologies in their role [37]. In addition to the lack of opportunity, lack of passion constitutes another hurdle in adopting digital education as many educators are unaware of the importance of having access to the new technologies of teaching and learning [38]. This may raise the question of whether the educators are ready enough and making themselves relevant in implementing education 4.0. A case study of applying education 4.0 in Malaysia revealed the problems faced by the HE Institution in teaching and learning based on the relationship between people, technology, and the environment [39]. The study identified several problems including the risk of future changes associated with Industry 4.0 that made the university's management unable to develop a strategy based on job market and technology trends. In addition, University's administration is struggling to find consistency between Industry 4.0 technologies and the existing university curriculum. Furthermore, obtaining approval from the accreditation body for redesigning the existing curriculum and multi-disciplines courses represented another challenge.

5.2 Lack of Regulations

Technologies of IR 4.0 could have significant negative externalities in scientific research. The WEF Global Risks Report in 2017 revealed that experts viewed the technologies brought by IR 4.0, as especially worrying. Among the potential problems are biotechnologies used to make massively destructive weapons, nanotechnology which may have negative effects on the environment or human health, and geoengineering that used for tackling climate changes and could have unanticipated consequences that irreversibly damage ecosystems [40].

Scientific and technological pursuits require freedom to push boundaries, but governments should think more fundamentally about how to contextualize new capabilities with reflection on purpose and values that includes the well-being of society. For example, in a famous speech in 1945 after the first use of atomic bombs, physicist J. Robert Oppenheimer offered his perspective that the purpose of being a scientist is to learn and share knowledge because of its intrinsic value to humanity. Biotechnologies, neurotechnologies, brain science, and virtual and augmented reality devices are likely to be more than any other set of technologies that impose ethical challenges. These technologies will have a lasting impact on life as they change how we interface with the world and they are capable of crossing the boundaries of body and mind, enhancing our physical abilities. Hence, the world urgently needs new approaches to agile governance, which govern technologies in ways that serve the public interest, meet human needs, and ultimately serve global civilization [4].

5.3 Resistance to Change

To date, most educators strongly believe that the best teaching method is traditional education. As a result, a high level of resistance to change occurs among educators toward the use of technologies in the classroom [41]. Lawrence et al. focused on the strength and weaknesses of Education 4.0 in the Higher Education system of Malaysia. He found that the commonly observed problem among several people in Malaysian universities is a very high resistance to change. Most lecturers in Malaysian universities consider that Education 4.0 will disengage the physical contact between students and faculties away from regular classroom culture [42].

6 Conclusion

Industry 4.0 will have an impact not only on technology advancements but also on people in the workforce. As such, HE institutions must properly train students in order to better adapt to such changes. Since IR 4.0 is built-in advanced technologies such as big data, cloud computing, data analytics, artificial intelligence, and machine learning, all these subjects should be included and taught comprehensively in the education programs. The educational content, as well as the methods of skill development, should meet the requirements of a new generation of employees. Additionally, the innovation and entrepreneurship management skills of the students should be developed through the restructuring of university programs. Moreover, a new leadership mindset and governance are crucial to tackle the challenges associated with education 4.0 and harness the future benefits of IR 4.0 while protecting humanity from potential threats carried by technology.

References

1. Pfeiffer, S. (2017). The vision of "Industrie 4.0" in the making—A case of future Told, Tamed, and Traded. *NanoEthics, 11*(1), 107–121.
2. Kagermann, H., Wahlster, W., & Helbig, J. (2013). *Recommendations for implementing the strategic initiative INDUSTRIE 4.0*. Final report of the Industrie 4.0 working group. https://en.acatech.de/publication/recommendations-for-implementing-the-strategic-initiative-industrie-4-0-final-report-of-the-industrie-4-0-working-group/. Last accessed 26 June 2022.
3. Schwab, K. (2016). *The Fourth Industrial Revolution: What it means, how to respond*. https://www.weforum.org/agenda/2016/01/the-fourth-industrialrevolution-what-it-means-and-how-to-respond/. Last accessed 20 June 2022.
4. Schwab, K., & Davis, N. (2018). *Shaping the fourth industrial revolution*. WEF. The Crown Publishing Group.
5. Whittle, M., & Rampton, J. (2020). *Towards a 2030 vision on the future of Universities in Europe*. European Commission, Directorate-General for Research and Innovation. https://rring.eu/wp-content/uploads/2020/10/KI0320599ENN.en_.pdf. Last accessed 26 June 2022.
6. Miranda, J., Corella, C., Noguez, J., Molina-Espinosa, J., Ramírez-Montoya, M., Navarro-Tuch, S., Bustamante-Bello, M., & Rosas-Fernández, J. (2012). The core components of education 4.0 in higher education: Three case studies in engineering education. *Computers and Electrical Engineering, 93*(107278), 1–13.
7. Chan, R. (2016). Understanding the purpose of higher education: An analysis of the economic and social benefits for completing a college degree. *Journal of Education Policy, Planning and Administration (JEPPA), 6*(5), 1–40.
8. Lim, J., Jo, H., Zhang, B., & Park, J. (2021). Passive versus active: Frameworks of active learning for linking humans to machines. In *Proceedings of the Annual Meeting of the Cognitive Science Society* (p. 43). Retrieved from https://escholarship.org/uc/item/28q879nz
9. European Students' Union. (2015). *Overview on student – centered learning in higher education in Europe: research study*. Brussels. https://www.esu-online.org/wp-content/uploads/2016/07/Overview-on-Student-Centred-Learning-in-Higher-Education-in-Europe.pdf. Last accessed 26 June 2022.
10. UNESCO IESALC. (2020) *Towards universal access to higher education: international trends*. https://unesdoc.unesco.org/ark:/48223/pf0000375686/PDF/375686spa.pdf.multi. Last accessed 22 June 2022.
11. Kantrowitz, M. (2021). *Who Graduates from College? Who Doesn't?: How to Increase College Graduation Rates without Sacrificing College Access*. ISBN-13: 979-8491199389.
12. Unnikrishnan, A. (2016). Retrieved from: http://economictimes.indiatimes.com/brics-article/skill-development-for-industry-4-0/brics_show/54460851.cms
13. Boston Consulting Group (BCG). Man and Machine in Industry 4.0. (2015).
14. World Economic Forum. The future of jobs report. October 2020. https://www3.weforum.org/docs/WEF_Future_of_Jobs_2020.pdf. Last accessed 26 June 2022.
15. World Economic Forum. Schools of the Future: Defining New Models of Education for the Fourth Industrial Revolution. (2020). https://www.weforum.org/reports/schools-of-the-future-defining-new-models-of-education-for-the-fourth-industrial-revolution/. Last accessed 26 June 2022.
16. Eliot, C. L. (1869). *The New Education*. XXIII.
17. Atkeson, A., & Kehoe, P. J. (2007). Modeling the transition to a new economy: Lessons from two technological revolutions. *American Economic Review, 97*(1), 64–88.
18. Geiger, R. (2013). *The land-grant colleges and the reshaping of American higher education*. Routledge.
19. Statista. Educational attainment distribution in the United States from 1960 to 2020. (2021). https://www.statista.com/statistics/184260/educational-attainment-in-the-us/. Last accessed 26 June 2022.
20. Research and market. (2020). *Global Online Education Market*. ID: 4986759.

21. Pozin, I. (2012). *10 Startups changing the world and what we can learn from them*. https://www.forbes.com/sites/ilyapozin/2012/05/09/10-startups-changing-the-world-and-what-we-can-learn-from-them/?sh=78622493271b. Last accessed 26 June 2022.
22. Brears, L., MacIntyre, B., & Sullivan, G. (2011). Preparing teachers for the 21st century using PBL as an integrating strategy in science and technology education. *Design and Technology Education: An International Journal, 16*(1).
23. Endy, D. (2017). *Yale-NUS College STEM Innovation Conference*. http://steminnovation.sg/wpcontent/uploads/2017/06/Endy_Yale_NUS_STEM_v1.pdf
24. Zunin, V., & Mammino, L. (2015). *Worldwide trends in green chemistry education*. Royal Society of Chemistry.
25. Ustundag, A., & Cevikcan, E. (2018). *Industry 4.0: managing the digital transformation*. Springer.
26. Penprase, B., & Nardin, T. (2017). *Common Curriculum at Yale-NUS*. https://indd.adobe.com/view/b8748bf2-c7a6-4cef-a1e6-9a30c36bfe80. Last accessed 25 June 2022.
27. JISC. (2020). https://www.jisc.ac.uk/reports/the-future-of-assessment. Last accessed 26 June 2022.
28. World Economic Forum. The Future of Jobs: Employment, Skills and Workforce Strategy for the Fourth Industrial Revolution. (2016), https://www3.weforum.org/docs/WEF_Future_of_Jobs.pdf. Last accessed 26 June 2022.
29. Universities of the future. Industry 4.0 implications for higher education institutions. https://universitiesofthefuture.eu/wp-content/uploads/2019/02/State-of-Maturity_Report.pdf. Last accessed 26 June 2022.
30. European Union. (2019). https://digital-skills-jobs.europa.eu/en/latest/news/eu-launches-two-new-initiatives-foster-higher-education-sector. Last accessed 26 June 2022.
31. Nisa, F., & Adesti, K. (2019). *Evaluation of the INDOPED Project*.
32. Gleason, N. W. (2018). *Higher education in the era of the fourth industrial revolution*. Springer.
33. Organisation for economic co-operation and development. OECD Reviews of Innovation Policy. China. Synthesis Report. (2007). www.oecd.org/sti/inno/39177453.pdf. Last accessed 26 June 2022.
34. Ministry of Education China. Regulations for Ordinary Institutions of Higher Learning. (2017). http://www.moe.gov.cn/srcsite/A02/s5911/moe_621/201702/t20170216_296385.html. Last accessed 20 June 2022.
35. Ministry of Education China. Notice from the Ministry of Education regarding jobs and entrepreneurship for graduates from Ordinary Institutions of Higher Learning in 2016. (2015). http://www.moe.edu.cn/srcsite/A15/. Last accessed 20 Mar 2022.
36. Ministry of Trade and Industry Singapore. Report of the Committee on the Future Economy. (2017). https://www.mti.gov.sg/Resources/publications/Report-of-the-Committee-on-the-Future-Economy. Last accessed 26 June 2022.
37. JISC. (2019). https://www.jisc.ac.uk/reports/digital-experience-insights-survey-2019-staff-uk. Last accessed 26 June 2022.
38. Siddiqui, M. A. (2007). *Qualities of a dynamic educational institution in Naqsh* (Vol. 6(4), pp. 49–57).
39. Mokhtar, M. A., & Noordin, N. (2019). An exploratory study of industry 4.0 in Malaysia: A case study of higher education institution in Malaysia. *Indonesian Journal of Electrical Engineering and Computer Science, 16*(2), 978–987.
40. World Economic Forum Global Risks Report. (2017). https://www3.weforum.org/docs/GRR17_Report_web.pdf. Last accessed 26 June 2022.
41. Abraham, O., & Amadi, R. O. (2016). E-education: Changing the mindsets of resistant and saboteur teachers. *Journal of Education and Practice, 7*, 122–126.
42. Lawrence, R., Ching, L. F., & Abdullah, H. (2019). Strengths and weaknesses of education 4.0 in the Higher Education Institution. *International Journal of Innovative Technology and Exploring Engineering (IJITEE), 9*(2S3), 2278–3075.

University-Industry Transformation and Convergence to Better Collaboration: Case Study in Turkish Food Sector

Ece İpekoğlu

1 Introduction

An efficient collaboration between university and industry is needed as a result of contemporary economic development challenges. It is crucial to enhance this collaboration to improve the education system, learning process, train students as the specialists of the future, make applied research, and transfer knowledge to potential professionals. Understanding of the goals and objectives of the collaboration between two parties is important to ensure successful university-industry projects. The aim of the university is to educate and train students for their future working life. In turn, industry has an expectation to employ highly qualified specialists who can use innovations and advanced technologies.

A study that starts on an intellectual basis within the university is completed at the end of a certain research and development process. It is reflected in the industry as a product, service, and technological development. At this point, it is most effective to transfer the knowledge produced in the academic field to the production process. This collaboration provides universities with a field of application for existing research. At the same time, it constitutes the driving force of national and regional development.

Speaking to the "Information Age" magazine on November 12, 2007, Toshiba's Chief Technology Officer Dr. Katsuhiko Yamashita explains the importance of cooperation with universities for industry:

> If you are a company operating in a country connected to the rest of the world, you are obliged to cooperate with universities.

E. İpekoğlu (✉)
Yaşar University, Bornova, Turkey
e-mail: ece.ipekoglu@yasar.edu.tr

In line with the words of a senior executive of one of the world's leading technology companies, knowledge is now a basic input. Both in national economic policy, in the development of sectors and technologies, as well as in company competitiveness and profitability. Parallel to the importance of knowledge, the importance of the intellectual capital that produces it is increasing. Therefore, in order to make a transformation in the higher education and research for the future, university and industry relationship should be also transformed and developed.

To begin with, the chapter attempts to highlight the importance of university-industry collaboration and introduce a case study in Turkey. Thus, it is aimed to capture points that will attract the attention of both academics and private sector employees and raise awareness in higher education with regard to transformation. Using deferent literature materials, the chapter gives the general overview on university-industry collaboration best practices which can be applied by developing countries. It highlights the experience on collaboration done in food sector in Turkey. Finally, it points the concluding remarks and advices that will lead to improve the collaboration.

2 Conceptual Framework: University-Industry Collaboration

The university-industry relationship actually has a direct relationship with science. The university-industry collaboration offers new prospects of research funds, real-world problems, and research challenges. The collaboration also creates innovation and provides national economic benefits.

The disconnections between scientific activity and industrial activity from time to time are due to the lack of university-industry cooperation that cannot be realized sufficiently. In order for the industry to produce technology, it is necessary to produce technology, and to produce technology, science must be produced. Moreover, there is a reciprocal relationship between education and industry. Just as industry has an impact on education, education also has an impact on industry. As Parsons and Shils [34] stated, it will be beneficial for educational institutions to have an educational-instructive structure and habit, by taking the initiative in the field of industry cooperation and guiding the industrialists with consistent and realistic approaches in this regard. It is the duty of educational institutions, as well as the business world, to train human resources as creative, proactive, able to find solutions to problems, productive, able to see the realities of the country, and follow the developments in the world. University-industry cooperation is as important for industrialized countries as it is for industrializing countries. It is essential for a country to be in good cooperation with the universities, which are the source of trained manpower and knowledge, and the industry, which is the source of finance, necessary for the industrialization of a country [41].

In this context, research and development is one of the main functions. Only in this way can the industry change its methods and tools and constantly offer new products to the market. This is only directly proportional to the intensity of research and development activities. Today, when countries are in the process of integration with the outside world, the level they have in science, technology, and well-trained manpower is very important. R&D studies and university-industry cooperation should be given due importance in order for the developments in this field to progress further.

University-industry collaboration have been discussed in the previous literature. Collaboration is regarded as resource, personnel, and information exchange with research projects, training or consultation partnerships [3–5]. Collaborating is an opportunity that university and industry join efforts and take an active role in preparing the next workforce to improve the process of education, to improve their reputation, and to obtain access to more research data. In another study on university-industry cooperation, they argued that the aim of academics is to discover and learn scientifically new information by using new and different applications, instead of making financial profit [38]. Wright et al. [39] evaluated university-industry cooperation in terms of economic development and stated that universities would gain importance in patent and license studies. University-industry collaboration is an innovative collaboration between entrepreneurial elements and universities or scientific research institutes [40].

As a result of this collaboration, the development of scientific and technological fields is ensured and it gives chance to students for interships and scientists for applying knowledge and technical tools within business circles such as production by putting their theoretical knowledge into practice [30]. When the definitions are evaluated, it is seen that the concept has many benefits such as the university's training and development of personnel for the industry, technology, R&D activities, and the support and financing of the university by the industry on various issues [20]. In some cases, the collaboration creates the partnership known as Industry-based learning (IBL) program in which universities send their students to industry partners for work experiences. The industry also benefits from this collaboration by assisting universities to prepare better equipped graduates to enter the workforce, since they are the potential employers.

In some studies, the role of the state in this relationship has been emphasized [9, 37]. Cohen et al. [11] and Fontana et al. [21] according to the results of their research have been determined that successful university-industry collaborations will mutually benefit both the university and the state, as the increase in sales revenues, productivity, and the number of patents granted. This cooperation increases the economic level of countries rapidly, improves the organizational learning ability of companies, and raises their innovation performance to higher levels [32]. On the basis of university-industry cooperation, there is the idea that such joint ventures will increase the R&D activities for the industry and the competitive power of the country, and also strengthen innovation.

Etzkowitz and Leydesdorff [17] proposed a triple helix model that includes university-industry and government relations in the local and national innovation cooperation process. It is a model in which there are dynamic tripartite relationships such as academic entrepreneurship, strategic alliances between companies, public-university-company research cooperation, joint use of facilities, and these relationships turn into intermediary institutions, networks, and creative organizations. The world is now shifting to this model in which three actors overlap their roles, solidarity and cooperation, and continuous communication is effective. It has been explained as the necessity of giving universities a more prominent role for innovation and economic development in the Information Society and the need for hybridization of this tripartite structure [18].

It has been in the last 30–40 years that University-Industry cooperation has gradually gained importance and emerged in different shapes and forms. However, cooperation; it has been at the center of many complex and difficult system analyzes and discussions of different dimensions, from necessary legal processes to cultural change, from national policies to institutional structures. While it takes a long time to design suitable models for successful practices and to provide the requirements, some conflicts between what universities are expected to do in these relationship processes and their actual functions have brought many different views to the agenda [26].

The relevant literature mostly emphasizes R&D studies and economic development purposes. However, the size of the expert to be trained should also be revealed. People who are university students today should be trained as qualified personnel for the industry in the future. In this context, it should be emphasized that students should be given support training programs on university-industry cooperation. The knowledge in the university should be transferred to the industry as soon as possible, and the technology existing in the industry should be transferred to the university. The education programs of the university should be directed toward the educational needs of the industry, university students should spend certain days of the week in the industry, and the knowledge they get from education should be brought to the business and the knowledge they get from the business to education. In order to have high-qualified human resources, which is the basic condition for success in a competitive environment, continuous and intensive training activities must be planned, developed, and applied in cooperation with industry and educational institutions [16]. Since developed countries understand this importance, they attach great importance to university-industry collaboration. In addition to increasing the work efficiency of the workforce through university-industry collaboration, it also assumes an adjusting and regulatory role for the labor market [23].

Last but not least, university-industry cooperation should not be a choice but a necessity for transformation of universities and research. This is not an issue to be overlooked by simple legislative barriers or formal practices. The parties should use a definite and clear will to create an effective and functional institutional cooperation structure.

2.1 The Role of University and Industry Collaboration on Innovation and Technology

In order to create new products and solutions, "innovation" is one of the key factors for industry. Innovation was defined for the first time by the economist and political scientist Joseph Schumpeter in his book written in 1911 and translated into English in 1934 as "the driving force of development" [15]. Innovation, according to the Oslo Manual is defined as "the application of new or significantly improved and renewed goods, services, processes or new marketing and organizational methods in various business lines, workplaces, and foreign relations" [33]. Making something new is defined as the process that ensures the widespread use and practice of new ideas [8].

Since innovation is the transformation of knowledge into economic and social benefit, it is the whole of technical, economic, and social processes. Desire for change, openness to innovation and entrepreneurial spirit bring success in innovation. As a result of innovation, companies whose productivity and profitability will increase and which will achieve high competitive power are developing, developing and gaining competitive advantage on a global scale [15].

It is tried to make a difference with the concept of innovation on the formation of the New World [25]. Carrying forward the developments in the phenomenon of innovation gives the countries superiority in all economic fields. In order for a country to gain international competence and successfully operate the phenomenon of innovation, the country must be able to operate interactively with its science and technology system (the system where applied research at universities with science centers and large public research laboratories are formed), that is, its production system and its science system. For this, too; it is important not to create any new unit or new institution, but to ensure that existing industrial organizations and universities, which are centers that produce science, work in cooperation.

Beside innovation, knowledge and technology transfer are also key benefits for the collaboration. By this way, both formal and informal knowledge and technology transfer could be realized. Formal transfer are patents, research papers, licensing agreements, and so on. Informal transfer, on the other hand, includes workshops, social networking, joint research projects, and qualified employees.

The aim of this chapter is to identify the reasons for collaboration between university and the industry by giving example from informal knowledge transfer side. The case, naming "Ask a Professional", is kind of a workshop and social networking program for students who will work in the food sector in the future. Qualitative research, particularly qualitative case study, is conducted for the study. Qualitative case study enables this chapter in exploration of a phenomenon within particular context as an example of university-industry collaboration "Ask a Professional" project from Turkey. By this approach, the chapter brings deeper understanding in order to reveal multiple facets of the phenomenon as Baxter and Jack [2] state.

3 Context of the Study

Many countries created policies to promote and sustain university-industry collaboration since 1980s [1]. The United States (U.S.) implemented pursuit of R & D in university and industry both originated nearly 125 years ago and have grown throughout the twentieth century. There has been a broad range of university-industry Cooperative Programs (U-ICP) in research and technology exchange in Canada as a result of debates about the lack of innovation in Canadian industry [35]. Moreover, the Science-to-Business Marketing Research Centre of Germany (S2BMRC) initiated a systematic study of cooperation among Higher Education Institutions (HEIs) in European Union (EU) countries and public and private organizations in Europe in 2010 and 2011 for the European Commission [13]. The UK is the best example of university and industry interactions with the contributions from UK university researchers and industry experts to problem solving in R&D projects [19].

The growing recognition is that Turkish academia should increase its main competitive advantages in both theoretical and practical area and so forth; it should adapt new approaches to foster innovation as a measure to compete efficiently in the global markets. In Turkey, university-industry collaboration is promoted with top-down strategy, not as a direct demand by industry [27]. Ideally, autonomous organizations can organically develop entrepreneurial culture with a bottom-up approach [22]. Therefore, academics support top-down push to start entrepreneurial activity would finally decrease the level of activity [36].

In Turkey, entrepreneurial behaviors and resource-seeking activities about higher education were started at the end of the 1980s. This was the time when Turkish universities began to participate in market-like behaviors through the technology transfer centers and techno centers [28]. After 1980s, Turkish universities are restructured and re-organized with regard to their new role as "entrepreneurship". Fifth university reform in Turkish Higher Education system in 1981 tried to adapt from Continental Europe to Anglo-Saxon-based model avoiding "Science for Science" mindset in order to encourage to overcome common barriers to entrepreneurial university model with the notion of "Science" in a knowledge-based economy [6]. However, these attempts are not enough to spread the entrepreneurial universities over the country. With regard to Technology Development Zones (TGB) by the 2001 regulation, engaging in technology push models is offered in weak entrepreneurial ecosystems which bring insufficient demand for innovation [10].

It is obvious that the universities in Turkey should transform their traditional structure into an entrepreneurial one [14]. There are some challenges in front of this transformation, such as resistance from faculty members, and the working conditions of the faculty members [12]. Turkey is in the first stage of the academic entrepreneurship process. Despite these challenges, there are some scholars who have already applied this model. However, the numbers are very limited. According to research by Okay [31], 43% of the faculty members did not participate in activities regarding university-industry collaboration. Bayrak and Muhsin [7] have a work including the faculty members and the industry representatives. According to this research, a lack of communication and reliability between two parties are the main obstacles for collaboration. Turkish universities will become more entrepreneurial and collaborative, but also, there is need for technology-intensive firms, in order to support on scientific directions, technologies, and for student placement in those firms for the future of the universities.

3.1 Case from Turkey: Ask a Professional

It will only be possible for universities to contribute to the industry by producing, monitoring, and evaluating innovations, and for businesses to train well-equipped manpower to be successful in global competitive conditions [7, p. 66]. The biggest obstacle to cooperation is that the two parties do not know each other well enough. Science produced at the university does not easily turn into technology. A large part of the current industrial sector does not know enough about which subjects and projects can be carried out at the university. Professionals who will provide this communication are needed.

Bringing the authorized experts in the enterprises to the status of lecturers and evaluating the students, or similarly, the academic staff working at the university to work for the industry can relieve this process. In general, there is a separation between education and research activities. While education is a more routine, self-repetitive system, research requires a more creative, different environment. Under the universities, research activities can be sustained with a more customer-oriented, project-based employment policy. In order to help small- and medium-sized industrialists, problem solving and consultancy activities in universities should also be encouraged. Likewise, interdisciplinary coordination should be encouraged. The concept of technology transfer means the transfer of information and technologies produced in universities and research centers to industry. This is the basis of university-industry collaboration. Strategic projects to produce knowledge can be produced at universities. Industry will also be the best places for applied research studies.

Emerging countries like Turkey reformed in the early 2010s, the academic promotion rules to offer new incentives as an addition to traditional criteria like publications. However, in many developing countries, institutional constraints may be an obstacle on the academic entrepreneurship according to Zuñiga [42]. For instance, in Thailand, some companies stated their interest in collaborating with universities, but they took a weak response from universities [24].

Having known all of these considerations, the project for students, who will work in the food sector in the future, was realized between a university and industrial company from Turkey. In this part of the chapter, how this project was implemented and its benefits to both the sector and students will be explained.

3.1.1 A Case Study: University-Industry Collaboration in Food Sector

Projects created in cooperation with university and industry in the food sector are very important because they bring together theoretical knowledge and "field" realities arising from practice. However, to bring together the forces of two "parties" with different disciplines and dynamics toward a common goal; two issues are important in the planning phase of the project. It was important to combine right project and right philosophy at the beginning of the project. Therefore, the correct positioning of the objectives of the project which can be called the "philosophy of the project" is firstly defined. The philosophy of this projects is to bring together future employees and current employees of food sector. From students' side, they would be realized where and in which conditions they would work in the future. Many students choose their higher education programs by chance. A few of them know what they want to do in the future. Therefore, "Ask a Professional" would give them a chance to understand whether they want to work in food sector or not. From industry perspective, they will recruit more deliberate candidate if this kind of workshop implement more.

Ask a professional emerged in these conditions and industry experts from food sector from a Turkish company and a scholar from a Turkish University created this project. Before, explaining the details of the project, it should be firstly mentioned "food worker/employee" concept.

Food sector differs from other part of the industry, because it has a production, quality, and R&D process that extends from raw materials to the end consumer; therefore, it requires cooperation from different disciplines. For this reason, in the university-industry collaborations to be realized in the field of food; it is essential to define the concept of "food worker". Food worker includes employees working in different parts of the food production chain. In this context, the employees of each unit have a function in the creation of the "food product" such as production, quality, R&D, planning, logistics, purchasing, sales and marketing, who are food workers.

To ensure success in cooperation projects, the ownership of the project by the food workers in the industry side; it is only possible if the food workers working in other parts have enough knowledge to show "empathy" for their functions. A food worker's attainment of a "multidisciplinary" consciousness is an issue that needs to be laid during the training s/he receives before starting to work in the workplace. Undoubtedly, a stakeholder, who is responsible for this awareness, is the university; because the only place where an employee who will work in the industry as a food worker will gain the notion of working with different occupational groups is the university or technical school where s/he is educated. As it can be seen, the university-industry cooperation is established when future food worker student starts while he/she is still studying. How the scholars responsible for educating the student will evolve into this functionality is a separate article and discussion topic. On the other hand, to train food workers is important for responsible food producers to invest in youth by taking part in a successful cooperation project.

A global company operating in the field of vegetable oil in Izmir started a project with the food engineering program from a university in Izmir, Turkey. The company has a consultant from this university, so they started the project partnership for years with Food Engineering program. In the last 2 years, they have carried this partnership even further and undersigned a project that they also call a social responsibility project within the scope of the university-industry cooperation. In this context, the project called "Ask a Professional" was started for students to improve themselves and their careers and to provide support to all young people who want to take part in the food industry in the future. This event should be also examined in terms of creating a culture of "food employee prone to cooperation projects".

"Ask a Professional" Project, which is planned 4 years ago with the company's Marketing Manager, started within the scope of the "New Techniques in Oil Refining" course that the project partner of the company gave to the third year students in the university's Food Engineering program. The project was first realized with the undergraduate and graduate students of Food Engineering program of the university, some parts of it in İzmir factories, and some parts at the university campus. Different departments such as R&D, Human Resources, Quality, Production, and Marketing provide students with information about the vegetable oil sector, the recruitment process, corporate life, and the fields they can work in. In this practice, which is carried out alternately between university and company every week, production, quality, R&D, marketing, sales, and human resources' unit managers were explaining a theoretical subject and, in this process, they conveyed the business life of the "food employee" by embellishing them with examples from their own careers. At the end of the event, the business was presenting the "Education Certificate" to the students.

So, what are the benefits of this program? In fact, as it is mentioned in the literature part, it is very useful to train students in the education process of the university to meet the needs of the industry in which they will take place as experts in the future, to understand the answer to the question of which department does what in an enterprise by spending time in the industrial facility in certain processes, it is very useful in transferring the knowledge obtained from the education to the enterprise in the future will be.

Collaborating universities is a win-win situation for both parties. University participation also motivates the companies for new research or adapting new technologies and acquiring laboratory equipments. By this way, they continue new projects with both theoretical and empirical research [29].

This event, which made an impression and received positive feedback from students; so, it will be advanced by adding the logistics unit and enriching the speaker staff with the participation of unit experts. The positive feedback provided the impetus to continue the process despite the pandemic. Although the event returns to online activity due to the developing pandemic; it expanded to include other programs related to food sector from different universities, such as the program of food technology, gastronomy, and cookery in different universities. After first groups, they continued the same program with other universities' food technology and gastronomy programs in Izmir.

3.1.2 Different Profession, Same Sector: Application Part of the Project

The company opened an application center (an application kitchen) which was a turning point. This kitchen, which was established to experience the performance of the oils produced under the company's brand in out-of-home consumption products with demo master chefs; it served as an important venue for the project, enabling to bring different professional groups together in the same kitchen. Thus, while the event turns into a practical food education; it has become a common platform where cooks, chefs, technicians, engineers, academics, and students come together.

This project for food employee candidates who come together for two educational purposes as the university's function of educating students who are inclined to a collaborative work culture becoming an exemplary activity that covers the social responsibility of the industrialist, who produces as a responsible producer, to train qualified personnel for the food sector.

As a part of the project, the company has recruited two food engineers, who attended this program when they were student, up to now. This is a beneficial for the company because these two employees have a preimpression about the vegetable fats and oil and also company culture, so their orientation process of the company was very fast and easy comparing their engineers. Moreover, after the beginning of the project, the university helps this company to make different research projects in line with the EU coming food regulations. This created really impact against competitors because they did not know much about new regulations which would be beneficial for gaining major clients.

4 Conclusion

Higher education institutions in Turkey, universities, are in the process of transformation, as a result of the global and neoliberal drives. For the future of the universities, the faculty members may keep their traditional roles as they are and continue in academia. However, it is essential for universities to create opportunities to be entrepreneurs and to collaborate the industry. Universities are main drivers for innovation and national and global economic development in the information age. On the other hand, industry can produce high-technology and innovative products based on academic researches, which brings about the concept of university-industry collaboration.

As it is defined in previous parts of the chapter, this collaboration is mutual, value-added, and experiential. Teaching and research have been considered as the main roles and responsibilities of the universities. In the global world, their third role is now defined as their collaboration with the industry as their stakeholders.

In Turkey, this collaboration between two parties has newly begun. Turkish universities started to participate in this kind of projects. Amongst recent reflections on university and industry collaboration, this chapter engages with the current transformation of higher education in Turkey with a case.

The implication of this chapter can be found as university-industry cooperation in higher education is not a choice but a necessity for the futures of the universities. This is not an issue to be overlooked by simple legislative barriers or formal practices. The parties should use a definite and clear will to create an effective and functional institutional cooperation structure. Technology Centers stand out as extremely efficient and indispensable applications that need to be developed.

References

1. Ayhan, A. (2002). *Dünden Bugüne Türkiye'de Bilim-Teknoloji ve Geleceğin Teknolojileri.* İstanbul.
2. Baxter, P., & Jack, S. (2008). Qualitative case study methodology: Study design and implementation for novice researchers. *The Qualitative Report, 13*, 544–559.
3. Brimble, P. (2007). Specific approaches to university-industry links of selected companies in Thailand and their relative effectiveness. In S. Yusuf & K. Nabeshima (Eds.), *How universities promote economic growth* (pp. 265–274). World Bank. Retrieved from: http://demo.netcommlabs.com/innovationgrid/pdf/How_Universities.pdf
4. Brundenius, C., Lundvall, B. A., & Sutz, J. (2009). The role of universities in innovation systems in developing countries: Developmental university systems–empirical, analytical and normative perspectives. In B. A. Lundvall, K. J. Joseph, C. Chaminade, & J. Vang (Eds.), *Handbook of innovation systems and developing countries* (pp. 311–325). Edward Elgar.
5. Correa, P., & Zuñiga, P. (2013). Public policies to Foster knowledge transfer from public research organizations. In *Innovation, technology, and entrepreneurship global practice, public policy brief.* World Bank.
6. Baskan, G. A. (2001). Türkiye de Yükseköğretimin Gelişimi. *Gazi Eğitim Fakültesi Dergisi, 21*(1).
7. Bayrak, S., & Muhsin, H. (2003). Öğretim Elemanları ve Sanayici Açısından Üniversite-Sanayi İşbirliğinin Değerlendirilmesi. *Manas Üniversitesi, Sosyal Bilimler Dergisi, 3*(5), 64–85.
8. Bjerregaard, T. (2009). Universities-industry collaboration strategies: A micro-level perspective. *European Journal of Innovation Management, 12*(2), 161–176.
9. Butcher, J., & Jeffrey, P. (2005). The use of bibliometric indicators to explore industry academia collaboration trends over time in the field of membrane use of water treatment. *Technovation, 25*, 1273–1280.
10. Clarysse, B., Wright, M., Lockett, A., Van de Velde, E., & Vohora, A. (2005). Spinning out new ventures: A typology of incubation strategies from European research institutions. *Journal of Business Venturing, 20*(2), 183–216.
11. Cohen, W. M., Nelson, R. R., & Walsh, J. P. (2002). Links and impacts: The influence of public research on industrial R&D. *Management Science, 48*(1), 1–23.
12. Collyer, F. M. (2015). Practices of conformity and resistance in the marketization of the university: Bourdieu, professionalism and academic capitalism. *Critical Studies in Education, 56*(3), 315–331.
13. Cunningham, J. A., & Link, A. N. (2014). *Fostering university industry R&D collaborations in European Union countries* (p. 19). The Whitaker Institute.
14. Dayioglu-Ocal, S. (2011). *Üniversiteler ile Teknokentlerdeki Eğitim Arge Şirketleri.*
15. Elçi, Ş. (2006). *İnovasyon Kalkınmanın ve Rekabetin Anahtarı.* Nova Yayınları.
16. Erkal, M. E. (2004). *Sosyoloji (Toplumbilim), 12.* Baskı, Der Yayınları, İstanbul.
17. Etzkowitz, H., & Leydesdorff, L. (1997). Universities and the Global Knowledge Economy: A Triple Helix of University-Industry-Government Relations. London: Pinter.
18. Etzkowitz, H., Webster, A., Gebhardt, C., & Terra, B. R. C. (2000). The future of the university and the university of the future: Evolution of ivory tower to entrepreneurial paradigm. *Research Policy, 29*, 313–330.

19. Fernandez, R. (2015). *Collaboration between universities and business in the UK* (p. 100). National Centre for Universities and Business, State of the Relationship Report.
20. Filik, Ü. B., & Ve Kurban, M. (2006). *Mühendislik Eğitiminde Üniversite-Sanayi İşbirliğinin Önemi ve Ar-ge Bilincinin Geliştirilmesi* (Elektrik-Elektronik-Bilgisayar Mühendislikleri Eğitimi, 3). Ulusal Sempozyumu.
21. Fontana, R., Geuna, A., & Matt, M. (2006). Factors affecting university-industry R&D projects: The importance of searching, screening and signalling. *Research Policy, 35*, 309–323.
22. Goldfarb, B., & Henrekson, M., (2003). Bottom-up versus top-down policies towards the commercialization of university intellectual property. *Research Policy, 32*(4), 639–658.
23. Gürbüz, E., & Ve Uçurum, E. T. (2012). Üniversite-Sanayi İşbirliğinin Geliştirilmesinde Ortak Araştırma Merkezi"nin Kurulmasına İlişkin Model Önerisi. *Niğde Üniversitesi İktisadi ve İdari Bilimler Fakültesi Dergisi, 5*(2), 12–36.
24. Hurt, E. (2012). The marketization of higher education. *College Literature, 39*(2), 121–132.
25. İnaç, H., Güner, Ü., & Sarisoy, S. (2007). Ekonomideki Değişen Devlet Anlayışı. *Akademik Bakış Uluslararası Hakemli Sosyal Bilimler E-Dergisi, 12*(3), 1–18.
26. Kiper, M. (2010). Dünyada ve Türkiye'de Üniversite-Sanayi Işbirliği ve Bu Kapsamda Üniversite-Sanayi Ortak Araştırma Merkezleri Programı (ÜSAMP) [*University Industry Cooperation in the World and in Turkey and University-Industry Joint Research Centers Research Program within this Context*]. Türkiye Teknoloji Geliştirme Vakfı.
27. Kremakova, M. (2016). The "new spirit of academic capitalism": Can scientists create generative critique from within? *Theory & Science, 1*, 27–51.
28. Küçükçirkin, M. (1990). Üniversite-Sanayi İşbirliği, Ülke Sanayi ve Ekonomi Açısından Önemi. *TOBB, Yayın, 158*(68), 1–18.
29. Lee, Y. S. (2000). The Sustainability Of University-Industry Research Collaboration: An Emprical Assessment. *Journal Of Technology Transfer, 25*, 111–133.
30. Odabaşi, A. Y., Helvacioglu, Ş., İnsel, M., & Helvacioglu, İ. H. (2010). Üniversite Sanayi İşbirliğinde Örnek Bir Model. *Gemi ve Deniz Teknolojisi Dergisi, 183*, 20–25.
31. Okay, Ş. (2009). Pamukkale Üniversitesi Öğretim Elemanlarinin Üniversite-Sanayi İşbirliği Çalişmalarina Bakişlari Üzerine Bir Alan Araştirmasi, Selçuk Üniversitesi ISSN 1302/6178. *Journal of Technical-Online Teknik Bilimler Meslek Yüksekokulu, 8*(2), 94–111.
32. Ömürbek, N. V., & Halici, Y. (2012). Üniversite Sanayi İşbirliği Çerçevesinde Antalya Teknokenti ile Göller Bölgesi Teknokenti Üzerine Bir Araştırma. *SDÜ, SBE Dergisi, 1*(15), 249–268.
33. Oslo Guide. (2006). Retrieved from https://www.tubitak.gov.tr/tubitak_content_files/BTYPD/kilavuzlar/Oslo_Manual_Third_Edition.pdf. February 2, 2022.
34. Parsons, T., & Shils, E. A. (Eds.). (1951). *Toward a general theory of action*. Harvard University Press.
35. Sá, C. M., & Litwin, J. (2011). University–industry research collaborations in Canada: The role of federal policy instruments. *Science and Public Policy, 38*(6), 425–435.
36. Scott, S. (2004). *Academic entrepreneurship university spin-offs and wealth creation*. Edward Elger Press.
37. Shane, S. (2004). Encouraging university entrepreneurship? The effect of the Bayh-Dole Act on University Patenting in the United States. *Journal of Business Venturing, 19*, 127–151.
38. Turk-Bicakci, L., & Brint, S. (2005). University-industry collaboration: Patterns of growth for low- and middle-level performers. *Higher Education, 49*. (½, 61–89.
39. Wright, M., Clarysse, B., Lockett, A. & Knockaert, M. (2008). Mid-Range Universities Linkages With Industry: Knowledge Types and The Role of Intermadiaries, *Research Policy, 37*, 1205–1223.
40. Wu, W. (2010). Higher Education Innovation in China, World Bank, Washington, DC
41. Yildiz, H. (2000). *Üniversite-Sanayi İşbirliği ve Kobi'ler, İ.Ü.* Adına TOSYÖV'e Hazırlanan Rapor.
42. Zuñiga, P. (2011). *The state of patenting at research institutions in developing countries: Policy approaches and practices* (WIPO Economic Research Working Papers 4). World Intellectual Property Organization. Retrieved from http://www.wipo.int/econ_stat/en/economics/pdf/WP4_Zuniga_final.pdf

Sustainable Development Goals Through Interdisciplinary Education: Common Core Curriculum at University of Hong Kong

Adrian LAM Man Ho

1 Introduction

In December 2002, the United Nations General Assembly declared *Decade of Education for Sustainable Development (2005 to 2014)* in its Resolution 57/254. The resolution aimed at mobilizing the diverse educational resources around the world to help create a more sustainable future. Therefore, all educational institutions were invited to incorporate the relevant principles, values, and practices of sustainable development into various aspects of education [57]. In September 2015, the United Nations General Assembly adopted *Transforming Our World: The 2030 Agenda for Sustainable Development* in its Resolution 70/1. Currently, this is deemed as one of the most universal and ambitious documents as it outlines the global plan of action for people, the planet, and prosperity working toward sustainability. The document highlights 17 Sustainable Development Goals (SDGs) as well as 169 specific targets and 232 unique indicators, which are pinpointing the series of major and common global challenges. In fact, these SDGs are extending and replacing the eight Millennium Development Goals (MDGs) and 21 targets adopted as early as 2000, which were once utilized to guide action for development to eradicate global hunger and poverty until 2015, but are lamentably associated with a number of uneven achievements and shortfalls [56]. Therefore, SDGs are now further expanding MDGs' scope, reach, and engagement in terms of their creation and implementation [23]. Most importantly, since the unprecedented and substantial outbreak of the coronavirus disease 2019 (COVID-19) in early 2020, the accomplishment of SDGs has become far more pressing yet simultaneously formidable,

A. LAM Man Ho (✉)
Common Core, University of Hong Kong,
Pokfulam, Hong Kong Special Administrative Region, Hong Kong
e-mail: u3519028@connect.hku.hk; lammanho@hku.hk

given that many of the existing exclusion, inequalities, and vulnerabilities around the world have been further revealed and exacerbated throughout the pandemic [59].

Among the 17 SDGs, SDG 4 emphasizes quality education, which reflects that education is not simply an integral part of sustainable development, but is also a critical factor in facilitating it. Meanwhile, it intersects with all the other 16 SDGs as education often contributes to their achievement [34]. Given the scale and significance of these SDGs, various actors in the community and the world are strongly encouraged to align their missions, efforts, and resources together. Ahmadein [1] mentions that the education sector is one of the very few actors that can help, promote, and contribute to achieving all the SDGs. University remains highly prominent in supporting the successful attainment of SDGs through the utilization of its expertize, proficiencies, and leadership. According to Leal Filho et al. [35], the possible contributions of universities to SDGs are manifold, which mainly include research and development; learning and teaching; governance and campus operations; as well as civic engagement and community outreach. By focusing on the aspect of learning and teaching, Albareda-Tiana et al. [3] comment that the complex and long-term sustainable development agenda calls for undergoing a paradigm shift in education, which goes beyond transforming institutional responsibility, but reorienting and reshaping the whole curriculum. Therefore, Chaleta et al. [13] propose the employment of an integrated and interdisciplinary approach to incorporating SDGs into the curriculum. In terms of integration, Chang and Lien [14] mention that General Education Curriculum (GEC) can afford holistic understanding and systems thinking when integrating sustainability into the overall university curriculum. While in terms of interdisciplinarity, Barth and Rieckmann [7] suggest that the successful incorporation of sustainability into the curriculum requires one to challenge traditional discipline-oriented and teacher-centered teaching, and simultaneously incorporate participatory and competence-oriented approaches. All these reveal possibilities brought by a university-wide and an interdisciplinary GEC in actualizing SDGs.

Hong Kong is no exception to the contemporary global trend of incorporating SDGs into universities. For instance, University of Hong Kong (HKU), which is the leading comprehensive university in Hong Kong, adopts interdisciplinarity as one of the fundamental pillars in its institutional strategy, also known as "3I+1 Is", apart from internationalization and innovation as well as impact. Regarding the aspect of interdisciplinarity, the creation of Common Core Curriculum (CCC) is one of its key highlights. All students will receive a broad and balanced GEC in their undergraduate education as one of their graduation requirements. CCC comes up with multipronged policies and practices in addressing SDGs, which mainly include mapping its comprehensive undergraduate courses taught by the ten faculties; offering a transdisciplinary minor around SDGs; organizing SDG activities and competitions; and promoting transdisciplinary research opportunities for students [18]. Although Blodgett and Feld [10] mention that there is never a one-size-fits-all manner to embed SDGs into the university curriculum, they are simultaneously aware that faculty is now constantly attempting to innovate and experiment in terms of learning and teaching sustainability by exploring issues and topics through multiple disciplinary lenses. Given all these, this chapter aims to utilize the specific case of HKU's CCC, especially in the aspect of the formal curriculum, to come up with some of the salient features and guiding principles for a university-wide interdisciplinary GEC.

The author structures this chapter into five major sections. This first section frames the contextual background of this study. The second section involves a short literature review of the relevant concepts about both interdisciplinary education and SDGs as well as GEC and HKU's CCC. The third section highlights the methodology for this study, while the fourth section presents the relevant findings and analysis. The last section sets out the implications and conclusion of this study, as well as reveals the future lines of research.

2 Literature Review

2.1 Interdisciplinary Education and Sustainable Development Goals

The attainment of all 17 SDGs can help secure a sustainable, peaceful, prosperous, and equitable life on Earth for both the contemporary and future generations [58]. All SDGs outline the major development challenges confronting humanity, and express the series of overall objectives and directions across various areas [66]. In particular, education which has become both a means and an end to SDGs, is now one of the top priorities among many universities. These universities have committed themselves concerning education to learn and teach the concepts, promote the relevant research studies, transform the campus environment and infrastructures, support efforts in the wider community, and engage and disseminate the results with international frameworks [55]. According to Prieto-Jiménez et al. [45], followed by the emergence of SDGs, there has been an increasing call that the series of associated problems and solutions should be addressed in an interdisciplinary, transversal, and holistic manner, which illustrates the inherently close linkage between both the concepts of SDGs and interdisciplinarity. Meanwhile, Sommier et al. [50] make a similar comment by highlighting that addressing sustainability requires the transcendence of the aims, perspectives, discourses, and methods of individual disciplines, given that it is linked to the ability to overcome complex and multifaceted problems with no distinct boundaries and straightforward solutions. Apparently, all of these are relevant to the idea of interdisciplinary education which involves the integration of different insights, values, and fields of knowledge. Students remain open and critical of different perspectives and interpretations, and propose a wide range of new and innovative ideas and solutions [33].

In the specific aspect of learning and teaching, many scholars have proposed various frameworks to allow the incorporation of SDGs, especially in the context of interdisciplinary education for cultivating sustainability practices [28]. According to Wiek et al. [63], five key competencies which help promote sustainability should be included in curricula and courses. First, for systems thinking, it involves investigating problems emerging and cutting across different domains, sectors, and scales. Second, for futures or anticipatory thinking, it involves anticipating how problems might evolve or occur over time with sustainable and desirable future visions, which involve the inertia patterns, path dependencies, triggering events, and alternative development pathways. Third, for values or normative thinking, it touches upon specifying, comparing, applying, reconciling, and negotiating values, principles, goals, and targets in various visioning, assessment, and evaluation processes. Fourth, for strategic or action-oriented thinking, it involves proposing and testing systemic interventions, transformational actions, and transition strategies, which leverage assets, mobilize resources, and coordinate stakeholders. Lastly, for collaboration or interpersonal thinking, in involves initiating, facilitating, and supporting various levels and modes of collaborative teamwork and stakeholder engagement. At the same time, the learning objectives of SDGs touch upon three dimensions as proposed by the United Nations Educational, Scientific, and Cultural Organization [58]. First, for cognitive dimension, it focuses on consolidating and enhancing the background knowledge and factual content of SDGs. Second, for socio-emotional dimension, it touches upon equipping the key competencies and values which facilitate and empower the application of SDGs. Lastly, for behavioral dimension, it emphasizes proposing innovative solutions and practical actions for applying SDGs to everyday life.

2.2 *General Education Curriculum Among Hong Kong Universities*

Since the 2012/2013 academic year, all the eight publicly funded universities in Hong Kong have substantially changed their undergraduate degree programs from three to four years. This additional year is accompanied by the inclusion of a GEC [64]. While each of the universities has subsequently come up with their own unique GEC structure and components, their common learning goals and outcomes touch upon broadening knowledge; critical thinking; interpersonal communication; problem-solving; civic responsibility; morality and ethics; global outlook; as well as life-long learning [21]. The concept of GEC is linked to the T-shaped educational model, whereas the vertical dimension represents one's specialized and disciplinary training, while the horizontal dimension is the broader experience attained through connecting and interacting with other disciplines [9].

GEC is widely regarded as one of the important avenues for universities to provide students with interdisciplinary learning opportunities, which can address the limits brought by the traditional focus on knowledge specialization and reductionist thinking within disciplinary silos [30]. According to Taylor et al. [54], these can be realized through developing a broad understanding and solving real-life problems in relation to sustainability. The overarching focus here is not merely about learning and producing relevant knowledge, skills, and dispositions, but also about goals, norms, and visions of transformation by framing from multiple perspectives and across various contexts. At the same time, GEC strives to offer all students a common set of capacities and capabilities in preparation for the diverse and unanticipated requirements in future studies, work, and life [62].

2.3 Common Core Curriculum at University of Hong Kong

In HKU, since 2009, CCC as the university-wide interdisciplinary curriculum was formally introduced for students to satisfy the requirements for GEC. The word "common" signifies that the formal curriculum and its many complements emphasize the commonality of human experience, while the word "core" focuses on the centrality of issues that are of deeply profound significance to humankind, as well as the core intellectual, social, and imaginative skills that all undergraduates should master upon their completion of study [17]. Under the active contribution of all ten faculties within the university, all academics offer students student-centered interdisciplinary learning experiences based on their specialized disciplinary expertize through their CCC courses. CCC is divided into four overarching Areas of Inquiry (AoIs), which include (1) Science, Technology, and Big Data (CCST); (2) Arts and Humanities (CCHU); (3) Global Issues (CCGL); as well as (4) China: Culture, State, and Society. Students generally need to take six CCC courses in total, one from each AoI and not more than two from any AoI. Nonetheless, the number of courses required as well as the year and semester in which they are taken vary across various programs. For instance, students who are coming from double degree programs only need to take four CCC courses, with one from each AoI [15].

To facilitate students' understanding and awareness of SDGs, many of the selected CCC courses are mapped against the 17 SDGs by their unique course nature and focus. All these can be identified by the SDG icons attached to each of the official course web pages. Meanwhile, five Common Core Clusters and Transdisciplinary Minors are offered for students to pursue an organized course of interdisciplinary study tailor-made based on their interests, and to address all the SDGs in ways that are modestly doable and make a positive difference. A cluster consists of four CCC courses while a minor consists of six CCC courses, which nonetheless are all drawn from the same thematic cluster. The courses under each cluster are listed for students' reference, and in all cases, there are many courses for them to choose from the list [18]. All these include Sustaining Cities, Cultures, and the Earth; The Quest for a Meaningful Life; Creative Arts; The Human Life Span; as well as Gender, Sexuality, and Diversity [16].

3 Research Methods

This study employs a single case embedded research design, whereas HKU's CCC is a single case together with a unit of analysis as the course description and learning outcomes of all CCC courses across the four AoIs [65]. In this study, there are three overarching research questions that guide the subsequent design and analysis as shown below:

1. What are the patterns of SDGs mapped against CCC courses?
2. What are the features of actualizing SDGs throughout the design of CCC?
3. How can CCC shed light on the underlying potentials of an interdisciplinary GEC in incorporating SDGs in the university curriculum setting?

To answer the three research questions, this study employed both quantitative content analysis and qualitative documentary analysis to understand the 270 CCC courses across all the four AoIs offered in the academic year of 2021/2022 as the research sample, which includes 75 CCST courses, 81 CCHU courses, 63 CCGL courses, and 51 CCCH courses. All these are publicly available written information available on CCC's official webpage. The author first utilized content analysis to count the distribution of SDGs and the number of SDGs for all CCC courses across the four AoIs. The author subsequently made use of open, axial, and selective coding to illustrate the general features of SDGs that emerged from the systematized database of course descriptions and intended learning outcomes of all CCC courses across the four AoIs. The author eventually gathered and interweaved the wide range of patterns and features together for attaining a holistic picture of how SDGs are incorporated into CCC.

There are several reasons for framing this study in this manner. First, according to Chang and Lien [14], an inventory of academic offerings is serving as the important baselines for comprehending both the hot and blind spots of the current courses, and identifying both the strengths and opportunities for further deepening and advancing the curriculum through interdisciplinary development and capacity-building. Second, a number of descriptive and empirical studies in the field (e.g., [26, 43]) have utilized SDGs as benchmarks to measure and assess the extent of integration of sustainability into the curriculum, which range from university to department level. Lastly, while curriculum can be analyzed through intended, implemented, and attained levels, the intended dimension always remains an ideal starting point of inquiry and a solid foundation for further comparisons among the above three dimensions [60]. Therefore, based on the author's best knowledge, given that there is highly inadequate or even limited study of SDGs in the context of Hong Kong's university curriculum, this study aims to illustrate the institutional orientation and desired purposes about SDGs at the macro level, with the specific case of HKU's CCC as an example of interdisciplinary GEC.

4 Findings and Analyses

4.1 Patterns of Sustainable Development Goals in Common Core Curriculum

Regarding the general distribution of SDGs for CCC courses across all the four AoIs, there are four SDGs mostly highlighted, including *Goal 11* (71 courses); *Goal 3* (55 courses), *Goal 16* (50 courses), and *Goal 9* (38 courses). Meanwhile, six SDGs remain particularly unattended, including *Goal 1* (3 courses), *Goal 2* (3 courses), *Goal 6* (6 courses), *Goal 14* (5 courses), *Goal 15* (3 courses), and *Goal 17* (5 courses). The framing of SDGs across CCC courses is focusing far more on holistic human development as well as governance and infrastructure, and relatively less on efficient and sustainable use of resources and dynamics of the Earth system. Situated in Hong Kong as a developed context with relatively well-off conditions, those dominated SDGs in CCC are representing the most important and prioritized ones for individuals to pursue further, while those unattended ones are not the most pressing and severe issues when compared to the developing counterparts. Upon closer examination, some goals like *Goals 14* and *15* are relatively more specific, which sometimes can be focused more by some courses originating from some disciplines. Meanwhile, some goals like *Goals 1, 2,* and *6* are touching upon fundamental human needs, which may be naturally included as part of the course discussion rather than the central focus of a course. The same applies to *Goal 17* when the attainment of many other goals is often embedded in the context of collaborative partnerships. Nonetheless, the inadequate emphasis on the latter two important dimensions reveals more efforts are needed for HKU to fully fulfill CCC's positioning by enriching the overall diversity of SDGs embedded into the curriculum.

While regarding the specific distribution of SDGs for CCC courses in each of the four AoIs, both similarities and differences are existing across them. First, for CCST, the three most common SDGs are *Goal 9* (17 courses), *Goal 3* (16 courses), and *Goal 11* (12 courses). This pattern is perhaps the most obvious one among all four AoIs. As many CCST courses are prepared by medicine, science, and engineering faculties and departments, their major focus is mainly looking into various aspects of science, technology, and innovation. These aspects are goals in and of themselves as key drivers for sustainability. Meanwhile, they are also central to the implementation of other goals as well as the support of translating abstract targets to concrete policies as well as measuring and evaluating impacts of sustainability. Second, for CCHU, the three most common SDGs are *Goal 3* (25 courses), *Goal 11* (20 courses), and *Goal 16* (16 courses). Given its broad and heavy emphasis on engaging with the sophisticated meaning of one's life through understanding, experience, and reflection, CCHU courses are contributed by almost all faculties and departments. Meanwhile, the attainment of well-being with a holistic and multifaceted nature, including the physical, mental, and social dimensions, requires very diverse contributions. Therefore, many CCHU courses are framed with individuals remaining as the center of discussion, such that considerations are made in terms of how to address all the external and internal issues which might influence maintaining and promoting well-being. Third, for CCGL, the three most common SDGs are *Goal 11* (20 courses), *Goal 16* (18 courses), and *Goal 10* (10 courses). The positioning of CCGL itself is extensive in both scale and intensity, which involves the investigation of a wide range of global issues by coming up with the causes, impacts, and solutions. One of the dominant discourses around the world is sustainability, which emphasizes striking a balance among the environmental, economic, and social dimensions. The attainment of such a goal will touch on issues and challenges across communities and cities as well as policies and institutions. Meanwhile, a problem-oriented perspective is employed in framing CCGL courses when inequality is deemed as the fundamental and major problem penetrating the contemporary world, whereas all other emerging issues are more like the manifestation, or simply by-products of such a much larger phenomenon. Lastly, for CCCH, the three most common SDGs are *Goal 11* (19 courses), *Goal 16* (8 courses), and *Goal 3* (8 courses). CCCH employs a highly similar approach in framing as CCGL as both are contextual rather than thematic as CCST and CCHU. This explains their shared commonalities in terms of the highlighted SDGs. Nonetheless, although the elements of the Chinese government, society, and culture are situated in an evolving global environment, the authentic context of China remains the central unit of analysis in CCCH. Therefore, given the prior developmental trajectory and the associated issues, there is now increasing attention to improving the well-being of Chinese individuals (Fig. 1).

Regarding the number of SDGs for CCC courses across all the four AoIs, most of the courses have mapped against two to three SDGs, whereas only a very few have mapped against four SDGs. Many academics generally recognize the importance and value of mapping SDGs in their CCC courses, which facilitate both the

Distribution of Sustainable Development Goals for Common Core Curriculum Courses across Four Areas of Inquiry (Compiled by the Author)

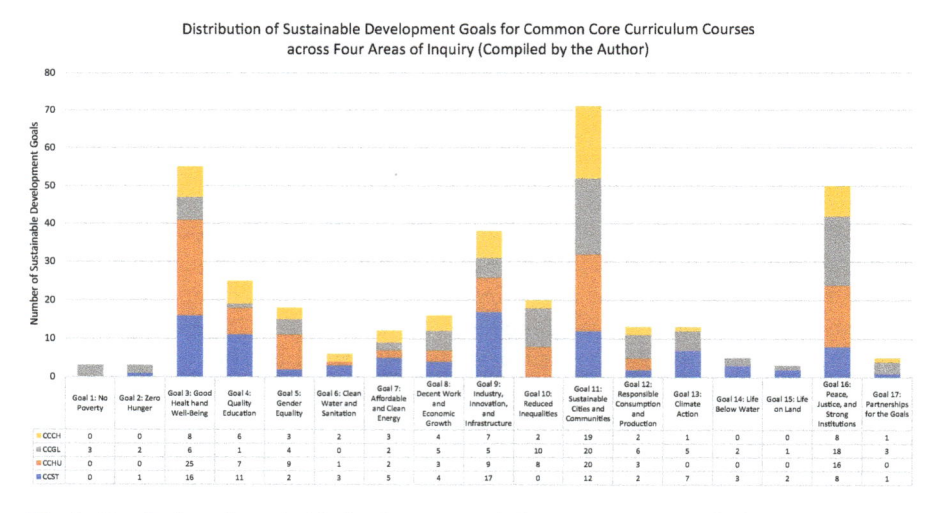

	Goal 1: No Poverty	Goal 2: Zero Hunger	Goal 3: Good Health and Well-Being	Goal 4: Quality Education	Goal 5: Gender Equality	Goal 6: Clean Water and Sanitation	Goal 7: Affordable and Clean Energy	Goal 8: Decent Work and Economic Growth	Goal 9: Industry, Innovation, and Infrastructure	Goal 10: Reduced Inequalities	Goal 11: Sustainable Cities and Communities	Goal 12: Responsible Consumption and Production	Goal 13: Climate Action	Goal 14: Life Below Water	Goal 15: Life on Land	Goal 16: Peace, Justice, and Strong Institutions	Goal 17: Partnerships for the Goals
CCCH	0	0	8	6	3	2	3	4	7	2	19	2	1	0	0	8	1
CCGL	3	2	6	1	4	0	2	5	5	10	20	6	5	2	1	18	3
CCHU	0	0	25	7	9	1	2	3	9	8	20	3	0	0	0	16	0
CCST	0	1	16	11	2	3	5	4	17	0	12	2	7	3	2	8	1

Fig. 1 Distribution of sustainable development goals for common core curriculum courses across four areas of inquiry. (Compiled by the author)

learning processes and outcomes. While it is theoretically possible for all courses to link to all SDGs in some sense, this selective representation out of the possible choices allows courses to show the SDG(s) that they deem as the most indicative one about their course nature and focus. It is worthwhile to mention that several courses did not explicitly and clearly indicate their linkages with SDGs. It is anticipated that these courses more or less will touch upon some elements about the SDGs as reflected by the subsequent analysis of course descriptions and intended learning outcomes. Nonetheless, to ensure alignment and coherence with other CCC courses with specific SDGs mapped, it is crucial for encouraging academics to make sense of their respective courses in the context of SDGs. Meanwhile, there could be further consideration of the primary and embedded SDGs in each CCC course, such that there could be a further idea of careful mapping and planning of these goals. Another side note is that although the SDGs generally mapped by them are roughly similar, CCCH has a relatively lower number of SDGs when compared to the other three. This is an expected observation as the total number of CCCH courses are relatively fewer when compared to the three other AoIs. Therefore, more efforts from the university are needed to invite academics to come up with their CCCH courses in the future (Fig. 2).

4.2 Features of Actualizing Sustainable Development Goals Through Common Core Curriculum

Regarding the major focuses highlighted by CCC courses across all the four AoIs, by looking into the course descriptions and course learning outcomes, there is generally a dual existence of both substantive content and proposed approaches in

Number of Sustainable Development Goals for Common Core Curriculum Courses across Four Areas of Inquiry (Compiled by the Author)

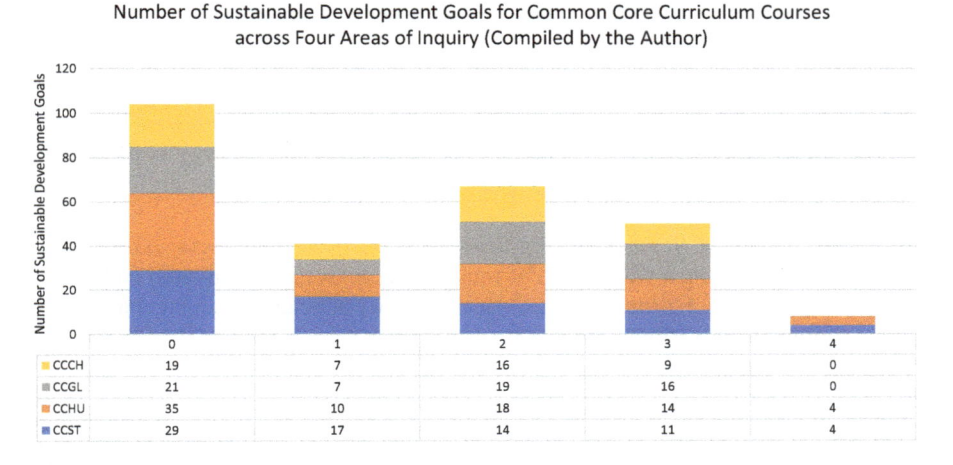

	0	1	2	3	4
CCCH	19	7	16	9	0
CCGL	21	7	19	16	0
CCHU	35	10	18	14	4
CCST	29	17	14	11	4

Fig. 2 Number of sustainable development goals for common core curriculum courses across four areas of inquiry. (Compiled by the author)

realizing SDGs in the respective courses. Interestingly, it seems that each AoI shows slightly different overarching goals and means. As students will take CCC courses from each of the four AoIs, this can facilitate them receive a holistic learning experience. Meanwhile, as students can freely choose their courses from each AoI under the distributional requirements, a central focus for each AoI helps guarantee the wide range of courses are aligned and coherent with one another.

4.2.1 Incorporation of Research Mentalities and Elements (CCST)

The thematic areas for CCST are primarily focusing on humans and life; health, energy, technology, and machine; as well as media. Students will mainly approach all of these through cultivating a critical understanding of the underlying impacts, issues, systems, and principles, as well as general skills and competencies for analysis. CCST aims to enhance students' levels of scientific and technological literacy given the knowledge, discourse, and advancement. Therefore, research is employed as the accessible strand to connect various elements. This is reflected by how different CCST courses are touching upon various parts of the systematic and logical trajectory of research and inquiry, such as formulating appropriate questions; coming up with relevant approaches; searching and collecting appropriate evidence; hypothesizing and testing possible solutions; as well as analyzing, presenting, and defending results and conclusions. Students are introduced to the research process and different important research skills [11]. Most crucially, academics will support such training throughout the course as this allows students to cultivate, consolidate, and deepen their capacities and capabilities gradually and progressively [31]. Throughout the entire process, students can benefit and develop a research mindset, and appreciate the transferability of these skills and mentalities [46]. According to Fayomi et al. [22], since SDGs are emerged out of an overarching intention of solving the generic problems confronting humanity, the fundamental role of research becomes pertinent to positively catalyzing the processes and procedures required to attain the reality of them.

- In the course "*CCST9022 How the Mass Media Depicts Science, Technology, and the Natural World*", students will be introduced to various scientific methods (i.e., observations, hypothesis, prediction, experiment, and theory) in the context of real-world issues, such as funding sources, control samples, and statistical calculations.
- In the course "*CCST9035 Making Sense of Science-Related Social Issues*", students will cultivate a wide range of transferable reasoning, analytical and evaluative skills through critical analysis of the impact of scientific and technological development on various real-world issues like equity, public health, and sociocultural practices.
- In the course "*CCST9017 Hidden Order in Daily Life: A Mathematical Perspective*", students will demonstrate analytical reasoning, formulate evidential and logical arguments, and present and communicate the coherent body of knowledge acquired.

The incorporation of research matters into CCST courses is particularly helpful for students to attain the most rewarding learning experience, given the fact that CCC courses are delivered by many active and experienced academics who are deeply engaged in the related research works, and are more familiar with those content base and understand the nuances thoroughly. Students can learn to embrace all ambiguity, complexity, uncertainty, and confusion, which are inevitable throughout the entire process of research and inquiry [53]. They are strongly encouraged to get rid of the flawless and all-knowing straightjacket as an essential trait of conducting research. Meanwhile, they can become confident in admitting their unavoidable knowledge and understanding gaps, and bear risks to try out new ways of continuous querying and active questioning, which can disagree or even challenge many of the well-established authorities and dogma [44]. All of these are essential qualities for preparing students to cope with the series of issues associated with SDGs that remain open-ended and unsolved as well as highly fickle and imperfectly defined. With the intellectual curiosity regarding how the larger world works and things become the way they are, students will assume the agency to further explore how to improve and change them correspondingly [38].

- In the course "*CCST9030 Forensic Science: Unmasking Evidence, Mysteries, and Crimes*", students will receive hands-on experience in collecting and analyzing several types of forensic materials, including hairs and fibers, fingerprints and shoeprints, soil samples, and drugs.
- In the course "*CCST9006 Chasing Biomedical Miracles: Promises and Perils*", students will apply their knowledge in authentic contexts, such as the human genome, organ transplantation, antibiotics, and cell communications.
- In the course "*CCST9045 The Science and Lore of Culinary Culture*", students will explore the series of concepts and theories behind the preparation, production, consumption, storage, and safety measures of everyday cooking and cuisines through laboratory demonstrations and practical sections.

4.2.2 Promotion of Deep Reflection and Assimilation (CCHU)

The thematic areas for CCHU are mainly focused on sex and gender; arts and cultures; as well as media and industries. Students will deal with these topics in the context of representations and portrayals; issues and conflicts; as well as ethics and moralities. CCHU targets crystallizing and examining students' understanding and experience of their times and life, on the foundation of a wide range of multifaceted problems manifested in contemporary society. Since any informed and thoughtful deliberation on one's own beliefs, values, experiences, and assumptions does not occur in a vacuum, the wide range of engaging and emotive issues offered by various CCHU courses become an effective medium for facilitating students' reflection and assimilation [49]. According to Taylor et al. [54], this does only limit students' ability to think critically about practices, opinions, and norms, but also their capacity to reflect on their own values, perceptions, and others, and understand external views. By making their mental models and sense-making active and conscious, students will become more aware of their positions, thoughts, feelings, and changes. This remains important as many of the ill-defined and ill-structured problems manifested by the interdisciplinary SDGs require one to undergo reflective thinking [48]. This overarching goal is generally attained through two major lines of developing the course, which include themes and topics, as well as genre and medium.

- In the course "*CCHU9005 Foods and Values*", students will examine philosophical issues about food and its relation to ethics, objectivity, and values, such as animal rights, world hunger, and agriculture genetic engineering.
- In the course "*CCHU9009 Moral Controversies in Contemporary Society*", students will investigate difficult issues of personal and public morality, such as euthanasia, assisted suicide, abortion, organ sales and donation, and prostitution.
- In the course "*CCHU9065 A Life Worth Living*", some of the world's major belief traditions are utilized to scaffold students' reflection, which include the very broad themes of what it means for life both to go well and to be led well, the reasons and motivations the tradition offers in its vision of a life worth living, resources each offers for attaining such a life, as well as what courses of action the traditions propose and advocate individuals are to do when they fail to live such a life.
- In the course "*CCHU9019 From Health to Well-Being*", students will develop a holistic and humanistic appreciation of the meaning of health in both a personal and societal context, which involves the analysis of qualitative and quantitative health information, literary and artistic works, as well as personal and collective introspection.
- In the course "*CCHU9016 The British Empire in Text and Image*", students will explain, explore, and appreciate the form and function of historical texts, the novel, poetry, travel writing, painting, sketch, and cartoon to examine the actual realities versus ideal representations of the British Empire.

As students are given numerous opportunities to work across disciplinary boundaries throughout various CCC courses, reflection remains important for them to better integrate theoretical and practical competencies, and raise awareness of the subtle and implicit dimensions [61]. With reflection involving both oneself and others, there are opportunities for students to take charge of their learning, and subsequently share and exchange ideas about all sorts of modifications and changes, as well as alternative and new ways of operating [27]. Students need to first undergo individual change and transformation, followed by changing the immediate community and larger world. Many CCHU courses often focus on those weak and underprivileged sectors, which are often struggling over inequality of opportunities and disadvantages as well as stigmatization and discrimination. While students in the future can make a difference through attaining SDGs, such difference should be grounded in clear thinking and sensitivity toward diversity and inclusion [29]. Therefore, CCHU courses are opening possibilities for them to consider how they can contribute their competences, interests, and abilities in initiating changes.

- In the course "*CCHU9007 Sexuality and Gender: Diversity and Society*", it utilizes debates to focus on people whose gender or sexuality are on the fringes of mainstream society, such as people who are gay, lesbian, bisexual, or asexual.
- In the course "*CCHU9010 Being Different: Understanding People with Disabilities*", it investigates the wide range of situations that people with disabilities commonly encounter in their everyday life situations, which include people with physical disabilities, and special educational needs, and mental illnesses.
- In the course "*CCHU9011 Social Divisions in Contemporary Societies*", it aims to enhance students' awareness of social divisions, and their implications for the distribution of resources and life chances in contemporary societies. This involves examining the socially constructed processes of social divisions among disadvantaged groups, analyzing their meanings from different theoretical perspectives, and identifying the ways to narrow social divides at the personal, societal, and policy levels.

4.2.3 Navigation of Systematic and Layered Analysis (CCGL)

The thematic areas for CCGL are mostly businesses and industries as well as institutions and politics. Students will investigate different general trends and changes; issues and crises; as well as processes and development. CCGL highlights thinking across levels and scales, which usually begins from systems, policies, and societies that are important and familiar as well as at times chronic and thorny from many individuals' perspectives. This is crucial to avoid students offering analyses and responses in a fragmented and piecemeal manner, especially when the series of issues facing humankind are influencing and interacting with each other as reflected by the wide range of SDGs. This applies to all sorts of considerations among individuals, such as causes and factors, decisions and actions, as well as impacts and outcomes, which involve co-benefits and trade-offs [42]. Students need to learn how to fit all isolated parts into a holistic and systematic picture of what the change exactly is and should attain. Meanwhile, this can shift their underlying mindset from linear and reductionist to circular and emerging [25].

- In the course "*CCGL9048 Global Crime and Injustice*", it explores the traditionally "local" arenas of criminological interest be located within a comparative angle, and articulated through transnational and global dimensions. Students will decipher the way things influence each other through feedback loops and flows within a system.
- In the course "*CCGL9040 Energy Futures, Globalization and Sustainability*", it states its overarching objective is to facilitate students to examine the relationships between energy, globalization, and sustainability at the global, regional, and local levels.
- In the course "*CCGL9049 Carbon, Money, and Lifestyle*", students are expected to describe and explain the multifaceted implications of climate change for individuals, society, and the economy, and demonstrate an awareness of the links between the above dimensions in the process of de-carbonization.

Throughout the layered analysis across CCGL courses, students will naturally and seamlessly conduct continuous conversations among as many academic disciplines, analytical perspectives, and guiding criteria, and formulate connections and relationships, as well as intersections and integrations. Students can eventually learn to think divergently and creatively for initiating positive impacts in their community and the world [51]. As students are working toward the attainment of SDGs through initiating various changes, changes in paradigms, beliefs, and practices will also be involved. On one hand, every single new decision can be followed by unanticipated implications or complications. On the other hand, the time and space needed to formulate fully researched decisions are almost impossible. The considerations, discussion, and deliberation will touch upon the roles, interests, and concerns of different decision-makers, stakeholders, and even interest groups, depending on the contexts of the system or society [24]. Meanwhile, single or multiple perspectives might be utilized, whereas the relative importance or priority of each perspective may also vary. There would also be common or competing claims or conflicting interests across groups, and tensions or dilemmas. Therefore, students should be capable of striking a dynamic balance between the dual poles and make optimal decisions at certain conjuncture, especially when constraints, costs, and trade-offs are inevitable [8].

- In the course "*CCGL9005 Poverty, Development, and the Next Generation: Challenges for a Global World*", students will examine the diverse strategies used by individuals and organizations, including nations, multilateral agencies, foundations, corporations, and individuals to reduce poverty and promote sustainable development.
- In the course "*CCGL9036 Dilemmas of Humanitarian Intervention*", it focuses on examining the dilemmas generated by great power politics, the tensions between state sovereignty and global humanitarian action, as well as the resource constraints in the contemporary world of potentially limitless need.
- In the course "*CCGL9062 Shaping our Health Across Cultures*", students are expected to critically analyze the tensions between various sectors within a society and government policy aims, and identify possible solutions to moderate them.

4.2.4 Facilitation of Critical Understanding and Interpretation (CCCH)

The thematic areas for CCCH are largely state and governance; culture and heritage; as well as history and tradition. Students will study various systems; policies; models; problems; issues; and challenges. CCCH aims to facilitate understanding of China both historically and in light of the issues currently faced by the country. Students need to grasp a comprehensive and detailed understanding through looking into various materials and cases, with the aid of theories, views, and perspectives, followed by exploring the social constructs and systemic structures of power and authority. Across the wide range of materials for students' exposure, some are either outdated or wrong, some are important and relevant, while some are even still under construction and emergence. The emphasis is on inspiring and guiding students to make articulations and integration, and consider, engage, and eventually formulate a cogent and thoughtful perspective from their own, on the foundation of the wide range of factual details and academic concepts incorporated and emerged throughout CCC courses [20]. Students can systematically articulate their beliefs and values, to explain how and why they have come to hold the views they do, and consider carefully to perspectives that might be different from their own [39]. According to Brookfield and Preskill [12], extensive and thorough discussions help one to unearth the diversity of underlying experiences and explore supposedly settled questions, and cultivate a fuller appreciation for the multiplicity of human experience and knowledge. The co-construction of learning between academics and students, and among students helps expand students' perspectives and possibilities in understanding and interpreting SDGs.

- In the course "*CCCH9010 Understanding China's Governance: Challenges and Prospects*", students will first by understanding the political economy of development through the gradualist reform, developmental state, and authoritarian resilience models, followed by analyzing the causes, scale, and dynamics of governance challenges, such as legitimacy, regulatory, distributive, and external ones.
- In the course "*CCCH9907 China in the Global Economy*", students will understand how China's rapid growth created new challenges, such as corruption, pollution, inequality, excessive debts, and over-capacity.
- In the course "*CCCH9025 Humanity and Nature in Chinese Thought*", it covers the ways various Classical Chinese Masters teachings impacted Chinese social ethics throughout its history, including Confucianism, Daoism, Mohism, and Chan (Zen) Buddhism. All had distinctive understandings and approaches to the relation between the human, social realm, and the realm of nature. Students can develop their view grounded on these culturally authentic concepts and patterns of thinking.

Although looking for the most updated sources is always essential, students often fail to consider the changes and continuities which lead to the existing situation with a certain context and background [36]. The employment of a chronological perspective allows students to look back and forth by tracing the historical development of the areas of SDGs on hand. This allows them to comprehend how the current situation is developed, what has been proposed and resisted, as well as why is there resistance, and would the contemporary environment is an optimal one for these resisted suggestions to be launched. Through looking into the rich history, students would realize that they often have attempted to raise and answer the most profound and enduring questions in human life, and have observed what are the gaps and limitations for them to overcome, as well as the mistakes and lessons to avoid, which are the underlying intention of proposing SDGs at the very beginning [2]. Meanwhile, this can help them to identify patterns that might otherwise be difficult to discern based on present events alone. Furthermore, students can compare and contrast multiple sources that represent differing points of view, which allow them to understand the complexity of the past and uncover potentially new perspectives [19].

- In the course "*CCCH9054 Mothering China: From the Womb to the Nation*", the focus is on how the idea of motherhood in China transformed from a personal experience to national duty and the question of how national leaders and social elites constructed, sustained, and altered the image of mothers between the late nineteenth century and now, a period marked by rapid socio-political changes, through the wide range of rich and authentic materials like texts, films, and adverts.
- In the course "*CCCH9044 Dunhuang and the Silk Road: Art, Culture and Trade*", students will make use of various analytic tools such as stylistic analysis, epigraphy, and archeological evidence to make sense of both the historical and contemporary Chinese culture and civilization from Dunhuang and the Silk Road.
- In the course "*CCCH9037 Chinese Mythology*", students are encouraged to consider the role of myths in both ancient times and contemporary modernizing society and the way the changing interpretation of specific mythological motifs can be analyzed as reflective of the wider changes in cultural values.

5 Implications and Conclusion

Universities always remain transformative agents and dynamic catalysts in terms of attaining the global agenda of sustainable development [47]. While many universities have already restructured their curriculum to include more interdisciplinary dynamics, there should be far more shifts in thinking, planning, and implementation among them. SDGs are essential for facilitating a fundamental and transformative shift in thinking, values, and actions across boundaries and individuals [40]. Nonetheless, both careful and detailed curriculum planning and design remain important to ensure these ambitious and vague goals can be included in a realistic and practical manner [14]. Therefore, such paradox justifies this study's overarching focus, which is utilizing a case study, namely HKU's CCC as a university-wide interdisciplinary GEC, to shed light on how academics can utilize SDGs as the broad, holistic, and flexible frameworks to frame, contextualize, and comprehend the series of issues, challenges, and complexities. This can avoid the tendency of possessing silo-thinking and introducing fragmented practices as those divides across the conventional specialized disciplines are naturally bridged in the authentic context of real-world themes as manifested by SDGs [6]. Meanwhile, one can bring all these wider issues into thinking about own core disciplines, distinctive experiences, as well as unique potentials, especially in terms of how to make progress and implement solutions toward SDGs realistically and strategically [35]. In particular, Lui and Lam [37] make a remarkable comment that the higher education sector has long been slow and stagnant in response to unprecedented and substantial crises with a series of abrupt and sweeping disruptions, given that the orthodox learning and teaching experience tends to be lagging behind rather than looking forward, or in other words, responding and surviving rather than foreseeing and thriving. As a result, this leads the overall approach remaining slow, unresponsive, and even resistant to the larger contextual changes. Therefore, the employment of visionary and futures-oriented SDGs as guidance for curriculum planning and design are particularly useful, especially in terms of addressing the gaps existing between learning and real-world situations. Most importantly, students will be facilitated to think, learn, and prepare for the complex and uncertain future ahead of them [4].

In this study, there are several implications regarding the incorporation of SDGs into university curriculum as reflected by HKU's CCC case. First, while comprehensive and balanced representation of SDGs remains as the central goal, some of them could be prioritized and highlighted in accordance with the unique contexts and features of the curriculum. An illustrative instance is how different AoIs have slightly different emphases, which all add up together to constitute a cohesive yet diverse learning experience for students. Meanwhile, the consideration of these contexts and features also maximizes the attainment of each of these SDGs [52]. Second, it remains crucial to include more SDGs and consider their interlinkages in the planning and designing stages of curriculum and courses [5]. Without such consideration of their alignment, coherence, and integration, these SDGs will simply become isolated and fragmented addition to the existing architecture, which bear no exact linkage to and impact on the underlying nature and focus of the curriculum and courses, and become far more tokenistic rather than genuine in its application. Third, since SDGs could potentially be criticized for being too abstract and generic, there are certainly in need of some realistic and concrete means to bridge, translate, and contextualize these goals with the curriculum and courses [32]. This leads to the emergence of a wide range of features actualizing the SDGs. While some of these means have long been existing in the conventional educational practices, the focus here should be reinforcing or even emphasizing their values and importance. In this study, there are four distinct and complementary ways for engaging SDGs in the space of learning and teaching, which include incorporating research mentalities and elements; promoting deep reflection and assimilation; navigating systematic and layered analysis; as well as facilitating critical understanding and interpretation. Lastly, SDGs are preparing one to become generative for the complex and changing dynamics in the future ahead. Therefore, SDGs should not simply be ritualized and adapted, but should serve as tools to stimulate thinking and foster innovation, especially when the road to sustainability remains complex and ongoing [41].

After all, this preliminary study aims to give a broad and general overview of how SDGs are integrated and embedded into HKU's CCC courses as a prototype of interdisciplinary GEC at universities. There are several lines for researchers to pursue for further studies in the future. First, given that this study only focuses on the ideal design, more empirical studies could be conducted to investigate the detailed implementation at the academic, classroom, and student levels, especially the extent of which the identified themes and features in this study are acted upon in the reality. Meanwhile, on the foundation of these baseline university-wide data, there could be more data collected from a wide range of sources, such as academics' reflection on practices and students' evidence of learning. Most crucially, there should also be attempts to operationalize and measure the exact impacts of SDGs at the curriculum level. Second, this study's findings can also be further analyzed against across other publicly funded universities in Hong Kong for obtaining a far more holistic picture, especially when there are many variations in terms of their introduction of GEC. This also helps fostering further discussion on the direction of desired changes in realizing SDGs at the university level. Third, while SDGs has become one of the top agendas in higher education, there could be more studies focusing on the ways in which various parts of the curriculum are interconnected and interdependent, such as the student-led research initiatives and out-of-classroom learning activities, as they all constitute a synergetic move toward the attainment of sustainability in the future. Therefore, more empirical studies could be conducted by focusing on the wide range of learning and teaching initiatives which embody elements related to SDGs.

References

1. Ahmadein, G. (2019). Obstacles and opportunities for achieving the SDGs at higher education institutions: A regional Arab perspective. In *Implementing the 2030 Agenda at Higher Education Institutions: Challenges and Responses* (p. 15). Global University Network for Innovation.
2. Akbaba, B. (2020). Investigation of chronological thinking skills of secondary school students and development of these skills based on grade level. *Egitim Ve Bilim-Education and Science, 45*, 215.
3. Albareda-Tiana, S., Vidal-Raméntol, S., & Fernández-Morilla, M. (2018). Implementing the sustainable development goals at University level. *International Journal of Sustainability in Higher Education, 19*, 473.
4. Alm, K., Beery, T. H., Eiblmeier, D., & Fahmy, T. (2022). Students' learning sustainability–implicit, explicit or non-existent: A case study approach on students' key competencies addressing the SDGs in HEI program. *International Journal of Sustainability in Higher Education, 23*, 60.
5. Al-Kuwari, M. M., Du, X., & Koç, M. (2021). Performance assessment in education for sustainable development: A case study of the Qatar education system. *Prospects*, 1–15.
6. Annan-Diab, F., & Molinari, C. (2017). Interdisciplinarity: Practical approach to advancing education for sustainability and for the Sustainable Development Goals. *The International Journal of Management Education, 15*(2), 73–83.
7. Barth, M., & Rieckmann, M. (2012). Academic staff development as a catalyst for curriculum change towards education for sustainable development: An output perspective. *Journal of Cleaner production, 26*, 28–36.
8. Blackwell, A. F., Wilson, L., Boulton, C., & Knell, J. (2009). *Radical innovation: Crossing knowledge boundaries with interdisciplinary teams* (No. UCAM-CL-TR-760). University of Cambridge, Computer Laboratory.
9. Blair, B. (2012). Elastic minds? Is the interdisciplinary/multidisciplinary curriculum equipping our students for the future: A case study. *Art, Design & Communication in Higher Education, 10*(1), 33–50.
10. Blodgett, D. M., & Feld, M. N. (2021). Teaching an interdisciplinary course in sustainable food systems: science and history meet in a world that works. International Journal of Sustainability in Higher Education, Vol. ahead-of print No. ahead-of-print. https://doi.org/10.1108/IJSHE-02-2020-0044 https://www.emerald.com/insight/content/doi/10.1108/IJSHE-02-2020-0044/full/html

11. Böttcher-Oschmann, F., Groß Ophoff, J., & Thiel, F. (2021, March). Preparing teacher training students for evidence-based practice promoting students' research competencies in research-learning projects. In *Frontiers in Education* (Vol. 6, p. 94). Frontiers.
12. Brookfield, S. D., & Preskill, S. (2012). *Discussion as a way of teaching: Tools and techniques for democratic classrooms*. Wiley.
13. Chaleta, E., Saraiva, M., Leal, F., Fialho, I., & Borralho, A. (2021). Higher education and sustainable development goals (SDG)—Potential contribution of the Undergraduate Courses of the School of Social Sciences of the University of Évora. *Sustainability, 13*(4), 1828.
14. Chang, Y. C., & Lien, H. L. (2020). Mapping course sustainability by embedding the SDGs inventory into the university curriculum: A Case Study from National University of Kaohsiung in Taiwan. *Sustainability, 12*(10), 4274.
15. Common Core Curriculum. (n.d.-a). *Common core for students*. Retrieved from https://commoncore.hku.hk/studentinfo/#requirement
16. Common Core Curriculum. (n.d.-b). *Common core clusters and transdisciplinary minors*. Retrieved from https://commoncore.hku.hk/studentinfo/#cctms
17. Common Core Curriculum. (n.d.-c). *Introduction to the common core – Engage, Experiment, Enjoy!*. Retrieved from https://commoncore.hku.hk/introduction/
18. Common Core Curriculum. (n.d.-d). *Sustainable development goals*. Retrieved from https://commoncore.hku.hk/sustainable-development-goals/
19. Cowgill, D. A., II, & Waring, S. M. (2017). Historical thinking: Analyzing student and teacher ability to analyze sources. *Journal of Social Studies Education Research, 8*(1), 115–145.
20. da Silva Almeida, L., & Rodrigues Franco, A. H. (2011). Critical thinking: Its relevance for education in a shifting society. *Revista de Psicología (PUCP), 29*, 175.
21. Eake, H. F. (2012). Curricular designs for general education at the UGC-supported universities in Hong Kong. In *General education and the development of global citizenship in Hong Kong, Taiwan and Mainland China* (pp. 121–136). Routledge.
22. Fayomi, O. S. I., Okokpujie, I. P., & Udo, M. (2018, September). . The role of research in attaining sustainable development goals. In *IOP conference series: Materials science and engineering* (Vol. 413, No. 1) (p. 012002). IOP Publishing.
23. Fisher, A., & Fukuda-Parr, S. (2019). Introduction—Data, knowledge, politics and localizing the SDGs. *Journal of Human Development and Capabilities, 20*(4), 375–385.
24. Fonseca, L. M., Domingues, J. P., & Dima, A. M. (2020). Mapping the sustainable development goals relationships. *Sustainability, 12*(8), 3359.
25. Fu, B., Zhang, J., Wang, S., & Zhao, W. (2020). Classification–coordination–collaboration: A systems approach for advancing Sustainable Development Goals. *National Science Review, 7*(5), 838–840.
26. Gómez-Martín, M. E., Gimenez-Carbo, E., Andrés-Doménech, I., & Pellicer, E. (2021). Boosting the sustainable development goals in a civil engineering bachelor degree program. *International Journal of Sustainability in Higher Education, 22*, 125.
27. Gray, D. E. (2007). Facilitating management learning: Developing critical reflection through reflective tools. *Management learning, 38*(5), 495–517.
28. Gual, C. (2019). Are Universities ready to have a real impact on achieving the sustainable development goals (SDGs)?. *Implementing the 2030 Agenda at higher education institutions: Challenges and responses* (p. 41).
29. Gupta, J., & Vegelin, C. (2016). Sustainable development goals and inclusive development. *International Environmental Agreements: Politics, Law and Economics, 16*(3), 433–448.
30. Hothem, T. (2013). Integrated general education and the extent of interdisciplinarity: The University of California–Merced's Core 1 Curriculum. *The Journal of General Education, 62*(2–3), 84–111.
31. Hughes, G. (2019). Developing student research capability for a 'post-truth' world: Three challenges for integrating research across taught programmes. *Teaching in Higher Education, 24*(3), 394–411.
32. Kioupi, V., & Voulvoulis, N. (2019). Education for sustainable development: A systemic framework for connecting the SDGs to educational outcomes. *Sustainability, 11*(21), 6104.

33. Klaassen, R. G. (2018). Interdisciplinary education: A case study. *European Journal of Engineering Education, 43*(6), 842–859.
34. Lawrence, A. W., Ihebuzor, N., & Lawrence, D. O. (2020). Comparative analysis of alignments between SDG16 and the other sustainable development goals. *International Business Research, 13*(10), 1–13.
35. Leal Filho, W., Shiel, C., Paço, A., Mifsud, M., Ávila, L. V., Brandli, L. L., et al. (2019). Sustainable Development Goals and sustainability teaching at universities: Falling behind or getting ahead of the pack? *Journal of Cleaner Production, 232*, 285–294.
36. Levstik, L. S., & Barton, K. C. (1996). 'They still use some of their past': Historical salience in elementary children's chronological thinking. *Journal of curriculum studies, 28*(5), 531–576.
37. Lui, W., & Lam, A. (2022). The critical engagement of higher education with future crises. *Dewey Studies., 5*(2), 345–362.
38. Lyytimäki, J., Salo, H., Lepenies, R., Büttner, L., & Mustajoki, J. (2020). Risks of producing and using indicators of sustainable development goals. *Sustainable development, 28*(6), 1528–1538.
39. Malloy, J. A., Tracy, K. N., Scales, R. Q., Menickelli, K., & Scales, W. D. (2020). It's not about being right: Developing argument through debate. *Journal of Literacy Research, 52*(1), 79–100.
40. Mataix, C., Lumbreras, J., Romero, S., Alméstar, M., Moreno, J., Mazorra, J., et al. (2020). Opportunity to foster urban innovation through universities: The case of Madrid. In *Sustainable development goals and institutions of higher education* (pp. 111–119). Springer.
41. Moallemi, E. A., Malekpour, S., Hadjikakou, M., Raven, R., Szetey, K., Ningrum, D., et al. (2020). Achieving the sustainable development goals requires transdisciplinary innovation at the local scale. *One Earth, 3*(3), 300–313.
42. Nilsson, M., Chisholm, E., Griggs, D., Howden-Chapman, P., McCollum, D., Messerli, P., et al. (2018). Mapping interactions between the sustainable development goals: Lessons learned and ways forward. *Sustainability Science, 13*(6), 1489–1503.
43. Pálsdóttir, A., & Jóhannsdóttir, L. (2021). Signs of the United Nations SDGs in University Curriculum: The Case of the University of Iceland. *Sustainability, 13*(16), 8958.
44. Peters, J., Le Cornu, R., & Collins, J. (2003). Towards constructivist teaching and learning. *A report on research conducted in conjunction with the learning to learn project, Australia* (pp. 1–20).
45. Prieto-Jiménez, E., López-Catalán, L., López-Catalán, B., & Domínguez-Fernández, G. (2021). Sustainable development goals and education: A bibliometric mapping analysis. *Sustainability, 13*(4), 2126.
46. Prince, M. J., Felder, R. M., & Brent, R. (2007). Does faculty research improve undergraduate teaching? An analysis of existing and potential synergies. *Journal of engineering education, 96*(4), 283–294.
47. Purcell, W. M., Henriksen, H., & Spengler, J. D. (2019). Universities as the engine of transformational sustainability toward delivering the sustainable development goals: "Living labs" for sustainability. *International Journal of Sustainability in Higher Education, 20*, 1343.
48. Richter, D. M., & Paretti, M. C. (2009). Identifying barriers to and outcomes of interdisciplinarity in the engineering classroom. *European Journal of Engineering Education, 34*(1), 29–45.
49. Schraw, G., & Dennison, R. S. (1994). Assessing metacognitive awareness. *Contemporary educational psychology, 19*(4), 460–475.
50. Sommier, M., Wang, Y., & Vasques, A. (2022). Transformative, interdisciplinary and intercultural learning for developing HEI students' sustainability-oriented competences: A case study. In *Environment, development and sustainability* (pp. 1–18). Oxford University Press.
51. Spelt, E. J., Biemans, H. J., Tobi, H., Luning, P. A., & Mulder, M. (2009). Teaching and learning in interdisciplinary higher education: A systematic review. *Educational Psychology Review, 21*(4), 365–378.
52. Stubbs, W., Ho, S. S., Abbonizio, J. K., Paxinos, S., & Bos, J. J. A. (2021). Addressing the SDGs through an integrated model of collaborative education. In *Handbook on teaching and learning for sustainable development*. Edward Elgar Publishing.

53. Tauritz, R. (2012). How to handle knowledge uncertainty: Learning and teaching in times of accelerating change. In *Learning for sustainability in times of accelerating change* (pp. 299–316). Wageningen Academic Publication.
54. Taylor, J., Jokela, S., Laine, M., Rajaniemi, J., Jokinen, P., Häikiö, L., & Lönnqvist, A. (2021). Learning and teaching interdisciplinary skills in sustainable urban development—The Case of Tampere University, Finland. *Sustainability, 13*(3), 1180.
55. United Nations. (2012). *Future we want—Outcome document.* Retrieved from https://sustainabledevelopment.un.org/futurewewant.html
56. United Nations. (2015). *Resolution adopted by the General Assembly on 25 September 2015, transforming our world: The 2030 Agenda for Sustainable Development.* The United Nations.
57. United Nations Educational, Scientific, and Cultural Organization. (2003). *United Nations decade of education for sustainable development (January 2005 – December 2014)* (Framework for a Draft International Implementation Scheme, July). United Nations Educational, Scientific, and Cultural Organization.
58. United Nations Educational, Scientific, and Cultural Organization. (2017). *Education for sustainable development goals: Learning objectives.* United Nations Educational, Scientific and Cultural Organization. Retrieved from https://unesdoc.unesco.org/ark:/48223/pf0000247444
59. Useh, U. (2021). Sustainable development goals as a framework for postgraduate future research following COVID-19 pandemic: A new norm for developing countries. *Higher Education for the Future, 8*(1), 123–132.
60. Van den Akker, J. (2004). Curriculum perspectives: An introduction. In *Curriculum landscapes and trends* (pp. 1–10). Springer.
61. Veine, S., Anderson, M. K., Andersen, N. H., Espenes, T. C., Søyland, T. B., Wallin, P., & Reams, J. (2020). Reflection as a core student learning activity in higher education-Insights from nearly two decades of academic development. *International Journal for Academic Development, 25*(2), 147–161.
62. Wehlburg, C. M. (2010). Integrated general education. *New directions for teaching and learning.*
63. Wiek, A., Bernstein, M. J., Foley, R. W., Cohen, M., Forrest, N., Kuzdas, C., Kay, B., & Keeler, L. W. (2015). Operationalising competencies in higher education for sustainable development. In *Routledge handbook of higher education for sustainable development* (pp. 265–284). Routledge.
64. Xing, J., Ng, P. S., & Cheng, C. Y. (Eds.). (2013). *General education and the development of global citizenship in Hong Kong, Taiwan and Mainland China: Not merely icing on the cake.* Routledge.
65. Yin, R. K. (2017). *Case study research and applications: Design and methods.* Sage.
66. Yuan, X., Yu, L., & Wu, H. (2021). Awareness of sustainable development goals among students from a Chinese senior high school. *Education Sciences, 11*(9), 458.

Implementing the Sustainable Development Goals (SDGs) in Higher Education Institutions: A Case Study from the American University of Beirut, Lebanon

Mirella Aoun ⓘ, Rami Elhusseini ⓘ, and Rabi Mohtar ⓘ

1 Introduction

1.1 The Scope and the Stakes

The Sustainable Development Goals (SDGs), adopted by all member states of the United Nations in 2015, describe a universal agenda that applies to all countries. UN members are urged to adopt and implement SDG policies in light of dwindling natural resources and impending climatic changes, as their palpable burden is increasingly felt on sustainable livelihoods and infrastructure worldwide [1, 2]. Higher education institutions (HEIs) were called on to take a leadership role, through research and educating students in relation to the goals and by inspiring engagement within their communities. While most universities now address environmental sustainability and/or sustainable development in some form, fewer universities have embedded an integrative or holistic approach to sustainability and the integration of the SDGs in both academic and applied domains. In this chapter, we will present a strategy for an integrative and holistic implementation of SDGs in HEIs, drawing on the case of the American University of Beirut (AUB), Lebanon. Through a higher vision, strategic direction, and collective efforts, AUB was able to surmount big barriers commonly identified in the literature as restricting the implementation of SDGs on university campuses [3, 4]. This was demonstrated by the

---

M. Aoun
Bishop's University, Sherbrooke, QC, Canada

R. Elhusseini (✉)
American University of Beirut, Beirut, Lebanon
e-mail: re52@aub.edu.lb

R. Mohtar
Texas A&M, College Station, TX, USA

© The Author(s), under exclusive license to Springer Nature Switzerland AG 2023
M. Ali S A Al-Maadeed et al. (eds.), *The Sustainable University of the Future*,
https://doi.org/10.1007/978-3-031-20186-8_12

significant leap made in its worldwide ranking against the global performance tables that assess universities integrating the United Nations' call on SDG partnerships [5]. The chapter will make the case for the implementation of SDGs in all the sphere of actions of a HEI, i.e., research and education in the age of Industry 4.0; sociocultural impact; operations; governance; and external leadership. It will describe the steps taken to formulate a strategic plan for the implementation of SDGs in all these domains by drawing on the case of AUB for inspiration. A strategy that will be described in five specific implementation components as follows.

2 Rallying Forces Around SDGs and Launching the Vision

2.1 The Importance of Senior Management's Commitment to SDGs

The importance of senior management's commitment to SDGs, its role in building up a vision that integrates them, its ability to effectively communicate the vision with campus community, and to rally the community around specific targets as illustrated in the case of AUB VITAL 2030 vision.

AUB adopts a system of shared governance where stakeholders partake in the constituency, and responsibility goes in tandem with accountability [6–8]. AUB's message and core values enshrine this institution in the fabric of Lebanese society which prides itself in being a culturally diverse and inclusive hub for the entire region. The long-lived partnership raised generations of leaders in influential positions, be they economic, agricultural, infrastructural, medical, or social, all imbued with AUB's liberal arts education and ethos [9]. The whirlwind of political transformations throughout Lebanon's modern history saw AUB stand firm, a flagship of social cohesion in addition to excellence in education. Tolerance, respect, and diversity in a polarized political environment came at great costs, however, requiring vigilance and untiring diplomacy. The parties did not always speak AUB's language, and many counterparts viewed dialogue as purposeless. Yet, AUB managed to include key players on its board of trustees, and through its dedication to shared governance, discourse, and a spirit of openness, it was able to align its priorities with those of Lebanon and the region [10]. This aspect was studied by Philip Zgheib of the Olayan School of Business (OSB) who analyzed the link between political networking potential and the impactful decision making of Lebanese expatriates worldwide. After that, his research sought to explain the origin of the Lebanese diaspora's involvement in international entrepreneurship, and in a third installment on the topic, Zgheib and his colleagues estimated the association between level of education and entrepreneurial impact [11–13]. Analyzing the answers of 264 Lebanese 1st generation entrepreneurs; Zgheib et al. found the level of education to be the most significantly associated factor with entrepreneurial-related variables, which include leadership and "entrepreneurial orientation" as coined by the authors.

Specifying the generational rank was critical because Lebanese expatriates have been settling abroad since the 17th century, making them an integral part of the societies they've adopted [14]. Albert Hourani for instance, who is a renowned reference on the topic frequently quoted in this research, is a 1st generation British Lebanese. He extensively researched Lebanese diaspora and taught at the AUB in 1938, yet he was born in Manchester England not Lebanon. His father, Fadlo Hourani, matriculated at the AUB a couple of decades before that [15]. This goes to show the enduring impact of AUB, and consequently, the impact of its graduates up to this day. Dr. Zgheib's coauthors included researchers from several renowned Arab universities, and included entrepreneurs, like Abdulrahim Kowatly, who eventually joined the OSB Faculty. Kowatly and Zgheib's 2013 project maps under SDG 8 "Decent Work and Economic Growth" according to Scopus. It clearly satisfies SDG 17 as well "Partnership toward the Goals", since the partnership of the coauthors coming from many institutions, led to a successful example of what the UN calls Education for Sustainable Development (ESD). ESD is described by climate and education experts Nhamo and Mjimba, as an essential characteristic of Sustainable Universities [16, 17]. Such visionary approaches parallel the international effort that produced the SDGs as formulated in 2015. Once the UN's Agenda 2030 is reassessed, it will undoubtedly include similar contributions to forecast the way forward. Starting 2019 AUB unveiled a new thematic approach called VITAL 2030 [18], forecasting a comprehensive plan to realign AUB's vision with its long-term goals and to benchmark its objectives.

The simultaneous cataclysms (political, economic, pandemical) endured by each stratum of the social fabric required quick reflex, and true to its pioneering second nature, AUB rose to the challenge. The result of several in-depth retreats spelled out five broad lines coupled with consequential long-term goals and specific stratagems with quantifiable elements to make sure the walk fits the talk. In VITAL, V stands for "Valuing AUB Community", I for "Integrating the Humanities", T for "Transforming the University Experience", A for "Advancing Research at AUB", L for "Life: Lifting the Quality of Health and Medicine in the Region."

The initiation of any research project at AUB closely follows the standard Plan-Do-Check-Act (PDCA) approach. Education experts Estévez and Chalmeta add the term "strategic planning" to the standard PDCA scheme. The same term which accurately describes the process is employed in the AUB's lexicon [19, 20]. In fact, one of the conditions research proposals must meet is compatibility with the faculty's strategic planning, which the research committee in every faculty ensures [21]. The research committee members are chosen from the faculty as part of their service to the community, an intrinsic aspect of shared governance at AUB. Strategic planning starts with the invitation of "directive" human resources to the AUB family, using Estévez and Chalmeta's characterization of successful contribution to methodology by an institution's Human Resources. Estévez and Chalmeta's work divides Human Resources' contributions into four vectors defined as a directive subdivision, a coordination group, facilitators, and action groups. Strategic planning starts with an invitation from the "directive" Human Resources to the institution's constituents. The directive action represented by upper management, is equivalent to the board of trustees (BOT) at AUB which instigates, contributes to, empowers, and validates the strategic planning. Through shared governance, the strategic planning which develops organically in each department, matures through faculty retreats, then back to the BOT table. The coordination group, represented by the board of deans (BOD), is headed by the provost. Through their authority delegated to the university librarian, they ensure that the generated data and dissemination of research follow the proper path, starting with the board of ethics or institutional review board (IRB) for instance, till the publication of results through the library's AUB-Scholar repository. The AUB-Scholar search engine will be discussed later to show the extent of SDGs related research. The facilitators' part is accomplished by the faculty members, partnering researchers, and advisors, but do not include funders or stakeholders, as by AUB's Grants and Contracts Department's regulations, in order to safeguard the integrity and independence of research. The action groups at AUB can be suitably represented by initiatives like AUB Innovation Park (AUB-iPark), which is an accelerator for new ideas started by students intending to launch a new concept into the marketplace. This will be discussed further in the next section. Stakeholders can be a part of the action groups, because once the research is published and the results are validated, conflict of interest subsides, and transmissibility/dissemination of best practices become the priority. What follows is a mapping exercise of VITAL 2030 to the 17 SDGs, listing actual practices/projects and echoing the gist of SDG 4's 7th goal, which characterizes education for sustainable development (ESD) as a catalyst for equity and equality, promotion of just and strong institutions, responsible consumption, and sustainable infrastructure, and most importantly partnerships for the goals. The first frame symbolized by VITAL's V, stands for valuing community and sharing AUB's core values. It naturally aligns with SDG 4 "Quality Education", SDG 5 "Gender Equality", SDG 10 "Reduced Inequalities", SDG 11 "Sustainable Communities", and SDG 16 "Just and Strong Institutions". Its objectives start with affirming and communicating AUB's unwavering commitment to accountability, cooperation, diversity, empathy, equity, integrity, respect, and transparency. VITAL 2030 sets forth concrete mechanisms to realize the goals of inclusivity without regard to economic status, recruiting the best academic prospects based on intellectual potential not financial abilities. Achieving these lofty goals can only take place when relevant and realistic funding is made available through endowments for research centers, accelerators and incubation hubs, and financial aid/scholarships. The sustainability of such plans is hence ensured, and strategic growth is streamlined with core values and ethos.

The I in VITAL is for "Integrating the Humanities" into purpose-oriented education and the 4th revolution; Industry 4.0. The benchmarking process in this area extends to all levels of AUB constituency, revealing the benefits of immediacy in shared governance. Calculating the success rate of a program's learning outcomes, could start with the class average in a midterm quiz and end up on a Board of Deans meeting agenda. E-learning, new media, data literacy, and technology imbue all the current disciplines. AUB has hybridized data into curricula and outreach, under the aegis of Industry 4.0. Digitizing the museum collection of the Archeology Department and the first editions of Arab Nahda manuscripts are good examples.

Offering 3-D printing workshops at the Interdisciplinary Design Practice Program takes the practice of concretizing the abstract to the next level [22–24].

AUB ScholarWorks can be used to further illustrate the practical ease in digitized research reporting [25]. A search for "SDG and Quality Education" will yield 128 results. Out of these results, more than 98% mention more than two SDGs and no less than 90% is dedicated research about enhancing sustainability as it directly relates to the 169 objectives of the 17 SDGs. Refining the search in the results page to include SDG and education for sustainable development (ESD) yields 39 results, the first is a policy brief by the Task Force on Climate Change, Sustainable Energy and Environment. It is a typical international collaboration published in partnership with the Department of Economics and spanning over three continents, several universities, and research centers [26]. It tackles the student initiatives relying on information and communication technologies (ICT) and the importance of incubation hubs like AUB-iPark to lead such projects to fruition and a spot in the market place. Enhancing the relevance of AUB as a center for innovation and research through its outreach activities, community service, and as accelerator can hence be measured as a quantifiable objective. The link to the Department of Economics following the search on AUB ScholarWorks shows the recent submissions of its faculty members to peer reviewed journals, showing SDG and sustainability-related research first. A further search of the Department of Economics' publications using "sustainability" for instance yields nine papers starting from 2010. Another search using "green" also yields nine publications, four in common with the first search.

T in VITAL stands for "Transforming the University Experience", it is the third pillar of the new thematic. It is validated through the practice of uncompromising transparency, a solid line of communication throughout the shared governance process, and the lived experience of student services and facilities, epitomized by the regenerating ease of campus life and immersed in respect, diversity, equity, and inclusion.

A: "Advancing Research at AUB" inevitably aligns the VITAL 2030 vision with SDGs 13 "Climate Action", 14 "Life below Water", 15 "Life on Land", and 17 "Partnerships for the Goals". The proximity of the campus to the sea allows for firsthand testimony of the urban carbon footprint unavoidable, with the biology department literally located by the shore. The Environment and Sustainable Development Unit (ESDU) at the Faculty of Agricultural and Food Sciences is yet another model of interdisciplinary and applied research. With international partnerships and endowments, this quintessential center combines outreach to state-of-the-art research where farmer, merchant, and researcher, in Lebanon and the region, advance the quest for knowledge along sustainable livelihoods [27].

Lastly, L in VITAL 2030 is the common denominator symbolized by "Life: Lifting the Quality of Health and Medicine in the Region", a theme synonymous with AUB's Medical Center and Faculty of Medicine. It aligns with SDG 3 "Good Health and Wellbeing", and it is validated by the world-class research that sets AUB apart. The development of a sustainable business model, giving all strata of Lebanese society and the region access to the latest clinical services relies on state-of-the-art data applied in medical records, distance learning, machine learning, and the promotion of a healthy lifestyle (Table 1).

Table 1 Mapping of the VITAL 2030 elements to SDGs

	VITAL 2030				
V	SDG 4 Quality Education	SDG 5 Gender Equality	SDG 10 Reduced Inequalities	SDG 11 Sustainable Cities and Communities	SDG 16 Peace, Justice and Strong Institutions
I	Industry 4.0				
T	Transforming University Experience				
A	SDG 13 Climate Action	SDG 14 Life Below Water	SDG 17 Partnerships for the Goals		
L	SDG 3 Good Health and Well-Being				

3 Connecting the University to SDG Networks

3.1 Relevant SDGs Networks and Active Contribution

This part emphasizes the importance of being effectively connected to relevant SDGs networks and how active contribution and mutual benefit advance toward SDGs implementation. Being connected stimulates motivation and fosters a culture of implementation in the different spheres of action of an HEI. Maintaining local, regional, and international ties facilitates education and research in support of the widely adopted SDGs. In September 2019, AUB inaugurated the Tala and Madiha Zein AUB Innovation Park (AUB iPark). In addition to immersing the students in latest Industry 4.0 tech and helping them "develop innovative ideas into profitable and scalable startups", AUB iPark centers on AUB's status as a hub of fact-based and deontology-prone entrepreneurship in Lebanon and the region [28]. The AUB iPark's incubation program heralds the digital revolution by initiating undergraduate students to the notion of expandable startups and validation of new ideas in order to build and accelerate a potential project then lead it to the market for execution and commercialization. The invitation extends to graduates, alumni, or researchers working with students. The gist of the initiative is to enjoin AUB students "from the heart of the startup ecosystem Beirut Digital District (BDD)" to convert an idea into a viable business by incubating it, accelerating the venture, connecting with industry specialists and mentors, and bringing together the creators of the idea with potential partners, talents, and investors, with all the projects listed in their database [28]. About the same time, AUB has partnered with Global Compact Network Lebanon (UN-GCNL) to offer students the chance of directly affecting the implementation of SDGs oriented policy. The Global Compact Network comprises around 70 local networks that are globally aiming to provide a "learning platform that facilitates heightened awareness, exchange of expert knowledge, and policy dialogue on sustainability and the 17 SDGs". The Lebanon network was started at the OSB Business school at AUB "with the ultimate goal of mobilizing a movement of sustainable companies and stakeholders to create a better Lebanon [29]." Both AUB and AUBMC partook in this collaboration becoming GCNL partners, and a Communication on Progress was published, as will be elaborated later in the chapter on the SDG reporting aspect. GCNL invested 100,000 USD and launched a call for proposals inviting AUB students to "incentivize and develop projects that have practical applications, with AUB increasingly being recognized as a hub for all SDG thinking and action across Lebanon." The successful applicants were invited to present their project at a dedicated conference at AUB, marking the inauguration of a partnership enduring the turbulence of the pandemic year. GCNL also collaborated with the Nature Conservancy Center at AUB (AUB-NCC) on the 2018 initiative International Biodiversity Day at AUB (IBDAA). IBDAA is an innovative approach to education with a focus on applied sciences; "IBDAA invites professors to enroll their classes in a competition to raise awareness of current environmental concerns by offering original and viable solutions" [30]. The GCNL led initiative in partnership with AUB took on "one of the Sustainable Development Goals set by the UN for 2030: 'Sustainable Cities and Communities' partnering with several Lebanese universities" under the AUBNCC umbrella.

In August 2021, AUB partnered with GCNL again to launch the call to join the "SDG Brain Lab Initiative", which included an incubation program, much like AUB iPark, culminating in a specialized training with professionals in the field amongst GCNL partners. The added value in this recent collaboration was the extension of the invitation to the medical school and nursing students through WEFRAH. WEFRAH is a Faculty of Agricultural and Food Sciences (FAFS) initiative that incorporates the foci of renewable resources and health into the Water-Energy-Food (WEF) Nexus and it aims to create a university-wide SDG-oriented approach, as will be discussed further in the chapter. The contributions of students in the domain of health bring in a fresh perspective to entrepreneurship that is grounded in the core values of AUB, as an HEI primarily committed to the educational and health service in Lebanon and the region. The invitation called on students to "join the SDG Brain Lab Program, an accelerator that aims to enable and nurture the change makers of tomorrow and empower the Lebanese youth between 18 and 24 years old in advancing the 17 Sustainable Development Goals of the UN 2030 Agenda on a national scale [31]." The innovative inclusion of future health professionals into the SDG fold brings a breath of fresh air to the field of healthcare because their AUB education centers these students around integrity and work ethic, allowing the future leaders and change makers to visualize a sustainable business model of health from a WEFRAH perspective. Such collaborations are a few examples of AUB centered initiatives around the SDGs theme. They can suitably serve as a model for university centered approaches tackling the most pressing concerns of the future generations, by providing the inevitable stakeholders a seat at the table, taking stock of decisions not just results. GCNL was launched in September 2015 and initially hosted at the American University of Beirut (AUB). In 2018, a legal entity was established for the sole purpose of hosting GCNL's Secretariat and the MOU with the UN Global Compact was updated reflecting this new hosting arrangement.

4 Promoting SDGs Initiatives on Campus

4.1 Implementing SDGs in a HEI as Holistic Approach

Promoting SDGs initiatives on campus: Individual initiatives targeting one or more SDGs exist in most HEIs. However, implementing SDGs in a HEI requires a collective, integrative, and holistic approach. Its success depends on how effective the HEI is in promoting and supporting campus-wide initiatives in research and education and developing cross-disciplinary team-based entrepreneurship and partnerships, reaching out to faculty, staff, and students in addressing global challenges. WEFRAH's AUB entry for University of Indonesia's GreenMetric call for posters "Universities, UI GreenMetric and SDGs in the Time of Pandemic" illustrates this part. The gist of the WEFRAH initiative consists in increasing synergy among various AUB departments, research centers, and policy units. Launched by the Faculty of Agricultural and Food Science (FAFS) to solicit engagement and multiply collaborations between AUB Faculties and Schools, it particularly promotes interdisciplinary partnership "by nurturing a bottom-up, participatory approach within AUB and between AUB and its external stakeholders." The methodology of WEFRAH stems from the nexus of Water-Energy-Food (WEF) in arid and semiarid climate, which our area has confronted since time immemorial or slightly after, congruent with historical/archeological accounts. Humanities are hence an indispensable partner whose contributions stand on equal footing with the financial, agronomical, or geopolitical perspectives. The innovative aspect in FAFS' approach was the inclusion of the Resources and Health fields to the WEF nexus. The proximity of other faculties, to name one factor, be it in terms of common specialties or casual scientific discussions, binds the AUB family in their quest for knowledge, providing fertile ground for WEFRAH. The initiative "opens venues for collaboration with other universities within Lebanon and its outcome and action-oriented approach links naturally and organically to the Sustainable Development Goals (SDGs) 2030 by presenting a contextualized, relevant approach to their achievement in the MENA region, using the AUB vehicle as a regional hub for innovations in water-energy-food-health nexus and interconnectedness in renewable resources in arid and semiarid regions [32]." A typical SDGs-related activity launched campus-wide and eliciting collaboration between faculty staff and students, was the project proposal addressing the salinity of the water well during the pandemic, in a remarkably dry year marked by shortages of water and power cuts. It was the AUB entry for the UI GreenMetric Sustainable Campus call for posters [33]. AUB's campus has been its flagship for the last five decades, but its water well became brackish due to decreased rainfall, and unsustainable consumption of the Beirut aquifer [34]. Extraction overload left the water table brackish and polluted. To mitigate the recently increased salinity, a plan for responsible consumption using less purchased fresh water was devised, with the help of experiential education and the innovative use of infrastructure. The movement restrictions of the pandemic were overcome through the efficient health system (streamlined vaccination program), and the educational facility (greenhouse/nursery labs) availing in-house services to the involved students [35]. AUB's campus is arguably one of the most memorable in the area, comprising a diverse variety of native and exotic flora. The in-house well sufficed during the past four decades but the new trend in climate change stifled that. Uncontrollable digging of wells due to lack of urban planning and exploding population numbers resulted in the transformation of most Beirut wells into brackish water reservoirs [36]. The physical plant department had launched a gradual replacement of salt-sensitive hedges with more salt-tolerant ones. Thanks to a comprehensive SDGs inclusive planning, a gradual transplantation of salt-tolerant species on campus to replace the cover plants requiring fresh irrigation water was already taking place [37]. However, the exponentially rising salinity recently exacerbated the problem. This led to an increased (and excessive) outsourcing of water, which kept the campus verdant but ran a hefty bill. By setting up nurseries using the expertize of the students in plant breeding, the program would be fast tracked while saving the cost of new saplings, and a separate brackish water system would be used to irrigate the new arrivals as they come out of the in-house nursery. The dynamic vaccination program, would give the students priority access to attend to the plant nurseries [38]. The initial campus-wide plan was to reduce freshwater use, then morphed into a proposal to fast track the multiplication and transplantation of high salt tolerance varieties. SDGs like "Life on Land", "Climate Action", and "Clean Water" buttress the theoretical core of the project, while others influence the methodology, specifically "Innovation in Infrastructure", and "Sustainable Urban Planning" both intrinsic to the AUB ethos of neighborhood partnership and improving quality of life. The proposal to replenish the campus with high salt tolerance species allows the sea flora, which had bestowed on the Lebanese coastline and Beirut their particular biodiversity since time immemorial, to reconquer a small space long lost to intensive urbanization and population density [39]. Beirut's coastline exemplifies the typical karstic propensity of the Lebanese terrain and is typified by the illustrious Raouche area's Pigeon rocks. The particular karst ecosystem and its accompanying biodiversity survived urbanization to a limited extent in the tiny part of the Raouche area called Dalieh, but it is increasingly bearing the brunt of "advancing city building" and overpopulation [40, 41]. AUB campus became the default protectorate due to its remaining patch of vegetation, defying the rabid real estate demand. Beirut has more trees on buildings than on land, and for the most part, its vegetation aims to denote a selective lifestyle rather than safeguarding the indigenous botanical patrimony. The advantage of AUB campus provides a much-needed natural extension and vital space for this ecosystem [42]. AUB is also situated in the natural continuation of the karst formation incline and is suited to act as shelter to many inhabitants of the Dalieh ecosystem, especially if the indigenous flora is allowed to thrive. Decreasing the effect of nonindigenous invasion favors wildlife conservation while fitting in a stricter freshwater use policy [43]. The reestablishment of the indigenous ecosystem may even extend to the sea floor, where algal diversity is intrinsic to life below water [44]. The reintroduction of endemic species and return of costal flora may end up saving more than water. The continuation of the habitat, downhill from the nearby ecosystem that Dalieh provides, can serve to protect many herptile and insect species that require the endemic seacoast flora to survive. Many indigenous and endemic amphibians and reptiles are currently endangered. Providing the suitable ecosystem which reestablishes their habitat would support the hidden lives of underground dwellers. The preserved underwater diversity may to be prove an essential cornerstone preserving the whole arch above, at an age of incrementally microbe resistance and dramatically receding microbiome [39, 45–47].

Another beneficial aspect of this initiative is the contribution to the "Quality Education" SDG by using the expertize of the Department of Agriculture of FAFS to train the students in hands-on plant breeding. The contextual nature of this exercise highlights the concept of solution-based thinking and addressing issues at the regional level, which the program has weaved into its vision and curriculum [48]. The WEFRAH initiative aims to influence policy and create research-based solutions. Based in FAFS, it focuses its vision through the lens of interdepartmental collaboration toward critical solutions in the context of arid and semiarid areas [32]. In this particular project, the collaboration with the Physical Plant Department allows for the optimal use of the irrigation system, by separating the brackish source and using the in-house well to irrigate without having to purchase freshwater [49]. Similar collaboration with the AUB vaccination program allows the fast-tracked vaccination of the students who would be working in the labs to breed the salt-tolerant species already found on campus, saving funds, promoting SDGs, and benefiting the students through experiential learning. The unsustainable practice, even if it is currently affordable, will contribute to depleting another aquifer, likely in the vicinity of the one no longer used for freshwater supply [50].

Another campus-wide project currently introduced is Solar Power. Elaborated on in the final part of this chapter, Dreaming the Future starts in the heart of the young through collegial discourse, experiential learning, exchange of ideas and expertize, and methodical grit.

5 Reporting and Increasing University SDGs Ranking

5.1 Tracking Progress in SDGs Implementation

Reporting and increasing the university ranking on SDGs: Tracking progress in SDGs implementation shows the extent of commitment of the university community toward the goals. Reporting on SDGs for the entire campus is quite a tedious task especially in large universities. This requires constant and regular tracking, the involvement of all the faculties, centers, and units, and the hard work of dedicated staff to collect and present an overall picture of sustainability on campus. The production of a sustainability report and the participation in international world rankings on SDGs help an HEI track its progress and measure the success of its strategic direction in implementing SDGs.

In 2019, the first SDG cohort of student-led initiatives in various SDG-related projects with high social impact presented their projects to stakeholders, faculty, students, and staff [51]. The event was the culmination of a campus-wide collaboration between AUB and Global Compact Network Lebanon, and led to the multiplication of similar projects, and the hybridization of new incentives to include elements of each previous success to advance participation from all likely contributors and allow new ideas to reach full potential, whether they have originated in the mind of a sophomore or a postdoc fellow. The SDG Brain Lab initiative by GCNL followed suit and included elements of AUB iPark then in the same spirit of collaboration, extended the invitation to all major Lebanese universities [28, 31]. The goal of these initiatives is matching the best idea with the most suitable audience through an incubation space which benefits from the most relevant expertize.

In 2021, FAFS' SDG initiative prepared a poster and video presentation to participate in the seventh International Workshop on UI GreenMetric World University Rankings (IWGM 2021) addressing the theme of "Universities, UI GreenMetric, and SDGs in the Time of Pandemic", held online on 25–26 August 2021 on Zoom and their YouTube Channel. Titled "Campus by the Sea: Adapting the Landscape to Evolving Salinity," it was included in the poster presentation session [52]. Surely enough, reporting the university's SDGs contributions, whether by incorporation into the curricula, or undertaking projects and partnerships toward these goals got noticed by the Times Higher Education World University Rankings (THE) [5, 53]. Even though 2020, the year of the pandemic, left its indelible mark on AUB, especially the Medical Center (AUBMC), reporting was relaunched and both a Sustainability Report for AUB and a Communication on Progress for AUBMC were published, detailing the institution's commitment to sustainable practices. It helped AUB jump from THE 300th rank to 87th between 2019 and 2021 [5, 53]. AUBMC's commitment to the 17 SDGs normally stresses its involvement in expected areas like health, safety, and the environment. Following the combined crises starting with the November 2019 revolution, the deeper involvement of AUBMC became evident. Whether through the lens of ER admissions or the solidarity of AUB community and its contribution to "Just and Strong Institutions" and "Equitable

Representation" SDGs, AUB's stakes in halting the country's downward spiral increased, as it became clear HEIs had a major part holding up the roof. Historically, "Good Health and Wellbeing" (SDG-3) had been the hallmark of AUBMC, due to its state-of- the-art healthcare service and steady involvement in social wellbeing providing the community with access to its services through the OPD clinics. "Climate Action", "Clean Water and Sanitation", and "Affordable Clean Energy" (SDGs 13, 6, 7) are inherent to AUBMC's normal functioning, mostly due to the vigilance of EHSRM Department (Environmental Health, Safety, and Risk Management). Starting from irrigation water use to the power plant, laboratories, pharmacy, kitchen, and laundry reaching the cafeteria and coffee shops, the emblems of recycling and sustainability signal the special awareness and effort dedicated to the environment. The existential challenge that led AUBMC to lay off an unprecedented number of its employees became a conduit to its involvement in several SDGs that were not directly addressed before. The ethical commitment to the community as a whole and to its staff, tied AUBMC to the emerging fight against poverty and even hunger. The devaluation of the Lebanese currency made the ghost of hunger a tragic reality, long after the collective memory of the 1916 Great Famine. SDG 1 "Fighting Poverty" hence became a priority, and although AUBMC was historically only involved as training grounds for "Quality Education" (SDG-4), it found itself committed to providing funding for education, through the extension of employees' education benefits to all the furloughs, to stave off another future crisis. The Beirut port explosion made it clear yet again that AUBMC's bond to Lebanon is primal. The reaction of faculty and staff in the face of this tragedy showed their mettle and elucidated their ties to the community. Economic sustainability, being sine qua non, led to a drastic review of the original expansion plan. Previous enthusiasm following Lebanon's brief economic spurt had led to what in hindsight was unplanned growth. The currency crisis forced some of the expansive plans to a grinding halt, leading the seemingly impervious institution to reckon with the notions of "Sustainable Growth, Reduced Inequalities, and Infrastructure" (SDGs 9, 11, 12). The "build it and they will come" optimistic approach had to evolve in light of the catastrophic countrywide situation, leading AUB to adopt a more prudent stance [54, 55]. Reporting the process not only benefits the institution by highlighting its SDG contributions, but it more importantly favors the transmission of experiential learning and benefits of hindsight.

6 Dreaming the Future

6.1 HEI Success in Implementing a Lasting Commitment to SDGs

Dreaming the future: An HEI's success in implementing a lasting commitment to SDGs can be demonstrated by its relentless dedication to make a difference, always going the extra mile to bring about positive change to society. An SDGs-focused HEI is well connected to its society and its needs. It can identify and bring-up consensus to support critical projects integrating key solutions to current global challenges. The challenges that compromise the health and wellbeing of future generations are specific to the environmental setting that each HEI shares with its community. This vantage point makes HEIs unique in their ability to locate and address these tasks.

AUB was built on a vision of Lebanon and region that can be best described as "dreaming the future". The fourth revolution only bolstered this ethos, as all the student-based initiatives show. When the dean of Agriculture, seven decades ago designated a 100 hectares plot in the midst of the Bekaa valley to found the 1950s state of art research and outreach facility, it embodied the industrial revolution. The center for Advancing Research, Enabling Communities (AREC) was the epitome of agricultural engineering, and represented the ideal of science in sheer practice, as hands-on as it gets [56]. The notion of fact-based and applied science was firmly planted, raising generations of leaders and pioneers of the field in Lebanon and the region. AREC introduced novel production techniques and innovations in agronomy, poultry, irrigation, crop production, etc. It was the unique approach to sustainable livelihoods that set it apart from the agricultural paradigm of the time. The idea of partnership between farmer and engineer, the notion of outreach made the dreamed future at reach. Even when the dream was interrupted during the civil war, the famine of 1915–1918 was never an option, due in great part to the extensive poultry production sector that AREC introduced to Lebanon [57]. Even when the dream was out of reach, it was surely protected. AREC now houses in its seed bank the collection of the International Center for Agricultural Research in the Dry Areas (ICARDA) since the war in Syria threatened the original seed bank located in Aleppo [58]. The introduction of no-till conservation agriculture (CA) was another milestone that further solidified the partnership of AUB toward the sustainable goals. The Millennium Development Goals at the time, leading to the current detailed version, permeated the interest in the first and foremost resources necessary for life, water, and soil. In July 2015 Milano, Italy during the World Expos' Universal Fair, Lebanon got a special recognition from the Expo organizers who awarded the Department of Agriculture at the American University of Beirut, a Best Food Security Practice Award for its project on Conservation Agriculture. The AUB research team had collaborated with the Ministry of Agriculture since 2007 introducing the practice of No-Till Conservation Agriculture to Lebanon. The pilot project ended in 2009 with an increase in No-Till from 40 to 500 hectares. By 2011 it had continued unabated, to reach 1100 ha, saving farmers around 300 $/ha/yr. in labor and land preparation. The Ministry of Agriculture has since formed a Committee on CA/No-Till, to actively promote the practice, with the aim of reaching 70.000 ha or a third of agricultural farmed land in Lebanon. It is a paradigm shift that will incrementally increase crop yield, cut labor costs, and reduce agrochemical seepage to underground water. It morphed the archetypal vision and creed of the farmer using state-of-the-art science and machinery to inculcate a new approach to farming based in fact-finding. Originally starting as a mere suggestion, the success of the no-till CA project is an example of the much-needed collaboration between farmer, student/researcher, and legislator. As many farmers felt ready to alter their perception of CA and adopt the No-Till system, the social perception of farmers is starting to follow lead, increasing the awareness of our need for partners toward a sustainable ecosystem, and a state of food security [59].

Currently, AUB leads another great vision of the future through the applied research on machine learning in the field of irrigation. The return to the source is exemplified in the constant concern with aridity, now a worldwide burden with the increasingly felt effects of climate change and modern production needs. The Internet of Things (IoT) holds the promise and the distant dream that a force beyond the horizon, like satellites, can help with the proper planning to overcome unplanned accidents like drought and floods. The advances in satellite-based estimates of irrigation requirements, and machine learning are currently the bulk of irrigation research activity done in the department of Agriculture. The ability to develop accurate algorithms using Google Earth Engine that would inform consequential programming related to agronomical practices epitomizes the IoT in practice [60, 61]. It entices the young just as the first mechanical horizontal feed mixer and distributor belts in a modern barn, in 1953, amazed the farmer and the agricultural student alike. Actualizing the vision of ubiquitous sustainability and conjuring a food secure world informed by fact-based theory and applied science start by teaching sustainability, incorporating it as a learning outcome in curricula. Many faculties include sustainability as a pillar in designing course syllabi, course learning outcomes, and program learning outcomes, so that novel ideas come full circle from the drawing board of a sophomore, to an alum's desk in parliament or the IMF, and back into the drawing board again for feedback and improvement [62, 63]. The aviation lighthouse sitting in front of AUB's lower campus, all but forgotten and dwarfed by current seafront luxury apartments, was a symbol of technological advances in communication when it was built. Transistor technology had surpassed previous lighthouse technologies, cutting edge as they once were, just as strong electrical projectors replaced optical systems or wood pyres and whale oil-based lamps before those. GPS technology and remote sensing advances have now replaced most of these compasses, and ongoing research is aiming to replace the fossil fuels running the engines they guided. Renewable energy like solar power is an indispensable resource, worth more than luxury flats, as green bonds are slowly outpacing classical mortgage, municipal bonds, or other debt securities [64–66]. It particularly rings true in Lebanon where their carbon footprint is felt on both climactic (micro) and economic (macro) sustainability levels [67–69]. The proposal to transform the Communication Tower near AUB, in collaboration with the Lebanese Ministry of Environment for instance and AUB's Center for Civic Engagement and Community Service, into a solar panel installation will not only recapture the innovative spark AUB invokes, but it will also harbor the dream of the future, a sustainable future and a legacy reflecting a deep understanding of the stakes.

Acknowledgments Special thanks to WEFRAH, the Faculty of Agricultural and Food Sciences (FAFS), and AUB leadership for helping promote this effort and the intrepid support during the darkest periods these past 2 years had to offer. Also, the office of Institutional Effectiveness and Decision Support (IEDS), the Physical Plant Department (PPD), AUB Medical Center, the office of Environmental Health, Safety, and Risk Management (EHSRM) for the continued collaboration and assistance during the harshest times of the pandemic and turmoil.

References

1. Schweikert, A., Chinowsky, P., Espinet, X., & Tarbert, M. (2014). Climate change and infrastructure impacts: Comparing the impact on roads in ten countries through 2100. *Procedia Engineering, 78*, 306–316.
2. Nhamo, L., Ndlela, B., Mpandeli, S., & Mabhaudhi, T. (2020). The water-energy-food nexus as an adaptation strategy for achieving sustainable livelihoods at a local level. *Sustainability, 12*(20), 8582.
3. Leal Filho, W., Shiel, C., Paço, A., Mifsud, M., Ávila, L. V., Brandli, L. L., Molthan-Hill, P., Pace, P., Azeiteiro, U. M., & Vargas, V. R. (2019). Sustainable Development Goals and sustainability teaching at universities: Falling behind or getting ahead of the pack? *Journal of Cleaner Production, 232*, 285–294.
4. Blanco-Portela, N., Benayas, J., & Lozano, R. (2018). Sustainability leaders' perceptions on the drivers for and the barriers to the integration of sustainability in Latin American higher education institutions. *Sustainability, 10*(8), 2954.
5. AUB among top universities in the world in delivering UN Sustainable Development Goals. https://www.aub.edu.lb/articles/Pages/Time_Higher_Education_Impact_Ranking.aspx
6. Hess, R. G., Jr. (2011). Slicing and dicing shared governance: In and around the numbers. *Nursing Administration Quarterly, 35*(3), 235–241.
7. Badr, L., Callaghan, D., Khoury, R., Mabsout, M., Meloy, J., Batakji, N., Nahas, A., Sadek, R., Saliba, N., & Sabra, W. (2011). The American University of Beirut.
8. Mouro, G., Tashjian, H., Daaboul, T., Kozman, K., Alwan, F., & Shamoun, A. (2011). On the scene: American University of Beirut Medical Center, Beirut, Lebanon. *Nursing Administration Quarterly, 35*(3), 219–226.
9. Anderson, B. S. (2011). *The American University of Beirut: Arab nationalism and liberal education*. University of Texas Press.
10. Bertelsen, R. G. (2016). The American University of Beirut: A case for studying universities in international politics. In *One hundred and fifty* (pp. 133–142). AUB Press.
11. Ahmed, Z. U., Zgheib, P. W., Carraher, S., & Kowatly, A. K. (2013). Public policy and expatriate entrepreneurs. *Journal of Entrepreneurship and Public Policy, 2*(1), 42–53.
12. Ahmed, Z. U., Zgheib, P. W., Kowatly, A. K., & Rhetts, P. (2012). The history of overseas Lebanese entrepreneurs operating worldwide. *Journal of Management History, 18*(3), 295–311.
13. Zgheib, P. W., & Kowatly, A. K. (2011). Autonomy, locus of control, and entrepreneurial orientation of Lebanese expatriates worldwide. *Journal of Small Business and Entrepreneurship, 24*(3), 345–360.
14. Hourani, A. H., & Shehadi, N. (1992). *The Lebanese and the world: A century of emigration*. Tauris Academic Studies.
15. Tripp, C. (2001). ABDULAZIZ A. AL-SUDAIRI: A vision of the Middle East: An intellectual biography of Albert Hourani. xiii, 221 pp. Oxford: The Centre for Lebanese Studies in association with IB Tauris, 1999.£ 25. *Bulletin of the School of Oriental and African Studies, 64*(2), 268–308.
16. Lukman, R., & Glavič, P. (2007). What are the key elements of a sustainable university? *Clean Technologies and Environmental Policy, 9*(2), 103–114.
17. Nhamo, G., & Mjimba, V. (2020). The context: SDGs and institutions of higher education. In *Sustainable development goals and institutions of higher education* (pp. 1–13). Springer.
18. Khuri, F. R. (2021). In Lebanon "it never rains but it pours"—How the American University of Beirut faced dangers and seized opportunities: Transforming medical education through multiple crises. *Faseb Bioadvances, 3*(9), 676–682.
19. Ferrer-Estévez, M., & Chalmeta, R. (2021). Integrating Sustainable Development Goals in educational institutions. *The International Journal of Management Education, 19*(2), 100494.
20. Reid, R. A., Koljonen, E. L., & Bruce Buell, J. (1999). The Deming Cycle provides a framework for managing environmentally responsible process improvements. *Quality Engineering, 12*(2), 199–209.

21. Procedures for Proposal Submission and Establishing Grants and/or Contracts. https://www.aub.edu.lb/ogc/Pages/procedural.aspx
22. Archaeological Museum. https://www.aub.edu.lb/museum_archeo/Pages/default.aspx
23. Archaeology Museum Virtual Tour. https://www.aub.edu.lb/museum_archeo/Tour/index.html
24. AUB Library Collections. https://www.aub.edu.lb/Libraries/News/Pages/Collections.aspx
25. AUB ScholarWorks. https://scholarworks.aub.edu.lb/
26. Holland, C., Mansouri, N., Bennett, D., Ni, H., & Dagher, L. (2021). *Youth transitions and transformations through ICT-enabled education for climate change and sustainable development*.
27. Environment and Sustainable Development Unit ESDU. https://www.aub.edu.lb/fafs/esdu/Pages/default.aspx
28. Tala and Madiha Zein AUB-Innovation Park. https://sites.aub.edu.lb/ipark/
29. Jamali, D., Samara, G., & Hossary, M. (2019). 12 corporate social responsibility and development. In *Business and development studies: Issues and perspectives* (p. 286).
30. IBDAA. https://www.aub.edu.lb/natureconservation/Pages/ibdaa.aspx
31. SDG Brain Lab. https://www.globalcompact-lebanon.com/hub/sdg-brain-lab-data-hub/
32. Water Energy Food Health Nexus, Renewable Resources Initiative. https://aub.edu.lb/fafs/wefrah/Pages/default.aspx
33. Elhusseini, R., & Battikha, G. (2021). *Campus by the sea: Adapting the landscape to evolving salinity*.
34. Saadeh, M., & Wakim, E. (2017). Deterioration of groundwater in Beirut due to seawater intrusion. *Journal of Geoscience and Environment Protection, 5*(11), 149.
35. El Beayni, N., Araj, G., Bizri, A. R., Khuri, N., & Shehabi, A. (2021). Available COVID-19 vaccine platforms: A roadmap to eclipsing the SARS-CoV-2 viral saga. *The International Arabic Journal of Antimicrobial Agents, 11*(1).
36. Lababidi, H., Shatila, A., & Acra, A. (2017). The progressive salination of groundwater in Beirut, Lebanon. In *Water and the environment* (pp. 279–284). CRC Press.
37. Makhzoumi, J. (2008). *Greening AUB's neighborhood I*.
38. Khuri, F. R. (2021). From cancer to COVID, Boston to Beirut. *Cancer, 127*(8), 1172–1173.
39. Kasparek, M. (2004). *The Mediterranean coast of Lebanon: Habitat for endangered fauna and flora: Results of a coastal survey in 2004*.
40. Mohsen, H., Raslan, R., & Bastawissi, I. E. (2019). The impact of changes in Beirut urban patterns on the microclimate: A review of urban policy and building regulations. *Architecture and Planning Journal (APJ), 25*(1), Article 2.
41. Raouche CCftPoDe. (2014). Landscape features of Dalieh. In *Beirut, Lebanon*.
42. Itani, M., Al Zein, M., Nasralla, N., & Talhouk, S. N. (2020). Biodiversity conservation in cities: Defining habitat analogues for plant species of conservation interest. *PLoS One, 15*(6), e0220355.
43. Norbury, G., Byrom, A., Pech, R., Smith, J., Clarke, D., Anderson, D., & Forrester, G. (2013). Invasive mammals and habitat modification interact to generate unforeseen outcomes for indigenous fauna. *Ecological Applications, 23*(7), 1707–1721.
44. Belous, H. K. O. (2014). Diversity investigation of the seaweeds growing on the Lebanese coast. *Journal of Marine Science Research & Development, 05*(01), 156.
45. Cuttelod, A., García, N., Malak, D. A., Temple, H. J., & Katariya, V. (2019). The Mediterranean: A biodiversity hotspot under threat. *Wildlife in a Changing World–An Analysis of the 2008 IUCN Red List of Threatened Species 2009, 89*, 9.
46. Cox, N., Chanson, J., & Stuart, S. (2006). *The status and distribution of reptiles and amphibians of the Mediterranean*. IUCN.
47. Çiçek, K., & Cumhuriyet, O. (2017). Amphibians and reptiles of the Mediterranean basin. In *Mediterranean identities: Environment, society, culture* (pp. 203–237). InTech.
48. Department of Agriculture, Faculty of Agricultural and Food Sciences at the American University of Beirut. https://www.aub.edu.lb/fafs/agri/Pages/default.aspx
49. The Physical Plant Department. https://www.aub.edu.lb/ppd/Pages/default.aspx

50. Bakalowicz, M. (2018). Coastal Karst groundwater in the Mediterranean: A resource to be preferably exploited onshore, not from Karst Submarine Springs. *Geosciences, 8*(7), 258.
51. FAFS hosts first SDG cohort. https://www.aub.edu.lb/fafs/news/Pages/2019_First-SDG-Cohort.aspx
52. Greenmetric, U. (2021). *The 7th international (virtual) workshop on UI GreenMetric world university rankings*. https://www.youtube.com/watch?v=EfLQzK-GPsM
53. AUB continues to climb in Times Higher Education rankings 2020. https://www.aub.edu.lb/articles/Pages/the-ranking-2020.aspx
54. AUB Sustainability Report 2021. https://www.unglobalcompact.org/participation/report/cop/create-and-submit/detail/459634
55. AUBMC Communication on Engagement. https://www.unglobalcompact.org/participation/report/cop/create-and-submit/detail/459601
56. Advancing Research Enabling Communities Center | AREC. https://www.aub.edu.lb/fafs/arec/Pages/default.aspx
57. Daghir, N. J. (2021). *Agriculture at AUB: A century of progress*. AUB Press.
58. Syrian Agricultural Experts Refresh Their Crop Management and Seed Production Skills. https://www.icarda.org/media/news/syrian-agricultural-experts-refresh-their-crop-management-and-seed-production-skills
59. Badran, A., Murad, S., Baydoun, E., & Daghir, N. (2017). *Water, energy & food sustainability in the middle east*. Springer.
60. Mourad, R., Jaafar, H., Anderson, M., & Gao, F. (2020). Assessment of leaf area index models using harmonized landsat and sentinel-2 surface reflectance data over a semi-arid irrigated landscape. *Remote Sensing, 12*(19), 3121.
61. Sujud, L., Jaafar, H., Hassan, M. A. H., & Zurayk, R. *Remote sensing applications: Society and environment*.
62. AUB Alumni. https://alumni.aub.edu.lb/s/1716/bp20/Interior.aspx?sid=1716&gid=2&pgid=403
63. Nehring, A. (2020). Naïve and informed views on the nature of scientific inquiry in large-scale assessments: Two sides of the same coin or different currencies? *Journal of Research in Science Teaching, 57*(4), 510–535.
64. Díaz, A., & Escribano, A. (2021). Sustainability premium in energy bonds. *Energy Economics, 95*, 105113.
65. Partridge, C., & Medda, F. R. (2020). The evolution of pricing performance of green municipal bonds. *Journal of Sustainable Finance & Investment, 10*(1), 44–64.
66. Banga, J. (2019). The green bond market: A potential source of climate finance for developing countries. *Journal of Sustainable Finance & Investment, 9*(1), 17–32.
67. Fakhreddine, B. B., & Faye, A. (2020). *A technical feasibility study of a hybrid wind/hydro power-system to provide firm power source and water for irrigation for the Uppermost Maten region–Lebanon*.
68. Bouri, E., & El Assad, J. (2016). The Lebanese electricity woes: An estimation of the economical costs of power interruptions. *Energies, 9*(8), 583.
69. Ahmad, A. (2021). Distributed energy cost recovery for a fragile utility: The case of Electricite du Liban. *Utilities Policy, 68*, 101138.

Index

A
ACT model, 88, 91–103
Adaptive leadership, 92, 98–100
American University of Beirut (AUB), 199–212
Artificial intelligence (AI), 17, 33, 101, 107–110, 112–114, 119, 120, 137, 149, 158, 159, 162
Augmented reality (AR), 108, 109, 112, 113, 149, 158, 162

C
Collaborative leadership, 102
Cooperative education, 202
CPD in higher education, 44–46, 48, 54–55, 57
CPD models, 48, 49
Curriculum of the future, 83, 95, 96, 102
Curriculum planning and design, 192, 193

D
Digital education, 157, 161
Disruptive technology, 114

E
Early education, 107–110
Education 4.0, xi, 120, 139, 150, 153, 155–162
Education for sustainable development, 5, 12, 43, 85, 86, 124, 177, 201–203
Eruptive technology, 25–38
Extended reality (XR), 112, 113

F
Fourth industrial revolution (IR 4.0), xi, 11, 95, 112, 119–122, 137–139, 149, 150, 152, 155, 160–162
Future education, 112, 155
Future University, x, xi, 25–27, 29, 32, 33, 37, 38, 123, 133–135, 137–139, 141, 145, 146

G
General education, 13, 178, 180–181

H
Higher education, v, vii, ix–xii, 3–6, 8–10, 12, 13, 16, 19, 20, 31, 36, 43–48, 52, 54–57, 63–81, 83–88, 100–102, 107, 111–114, 119, 120, 122, 124, 135–137, 139–141, 149, 151, 153–158, 162, 166, 170, 172, 174, 175, 192, 194, 209
Higher Education Institutions (HEIs), v, vii, ix–xi, xvi, 2–7, 12, 13, 19, 20, 25–36, 38, 46, 48, 52, 54, 56–58, 81, 83, 84, 86, 88, 90, 91, 93–95, 97, 100–103, 138, 139, 170, 174, 199, 206, 210
Hong Kong, xi, 178, 180–183, 194
Humanising education, 123–126, 128, 130
Humanities, xi, 33, 43, 65, 69, 71, 72, 80, 94, 97, 121, 125, 126, 130, 133–138, 140–146, 160, 162, 179, 181, 186, 191, 201, 202, 206

I

Industrial Revolution, 95, 107, 111, 120, 133, 137, 139, 150, 153–155, 159, 211
Industry 4.0, ix, xi, xv, 112, 113, 119, 139, 149, 150, 152, 155, 156, 161, 162, 200, 202, 204
Interdisciplinary, 7, 11, 31, 50, 94, 95, 101, 103, 107, 135, 140, 143–146, 150, 155, 171, 178–182, 188, 192, 193, 203, 206
Interdisciplinary curriculum, 155, 181

P

Praxis, 83, 91, 93, 94, 100, 102, 103
Predictive learning analytics (PLA), ix, 25–28, 30–38
Professional Development in Higher Education, 46, 52–54

R

Regenerative leadership system, 88, 91–97, 99, 102, 103
Reorienting higher education, 86, 102
Research, v, ix–xii, xv, xvi, 2, 3, 5–20, 26, 27, 29, 30, 37, 38, 44–48, 51–54, 56–58, 64, 66, 67, 74, 79, 86, 96, 100, 110, 113, 116, 120, 121, 128, 134–138, 140, 142–145, 149–151, 153, 154, 156, 159–161, 165–171, 173, 174, 178, 179, 182, 186–187, 193, 194, 199–204, 206, 211, 212
Research and higher education, 166
Respect and partnership, 129
Role of Industry, 174

S

SDG leadership labs, 99–100, 103
6G, 107, 108, 112, 114–116
Smart cities, x, 18, 108, 110, 111, 114, 116
Social sciences, xi, 93, 110, 133–138, 140–146, 160
Sustainability, ix, x, 2–20, 43–58, 63–66, 70, 75, 76, 80, 83, 85, 86, 88, 95, 96, 100–104, 110, 116, 123, 124, 128, 129, 138, 177–184, 189, 193, 194, 199, 202, 203, 205, 209, 210, 212
Sustainability in Higher Education, 44
Sustainable development era (SDE), 83–104
Sustainable Development Goals (SDGs), ix, xi, xii, xv, 2–20, 43, 45, 83, 85–88, 90, 94, 98–100, 102, 103, 124, 127, 128, 130, 177–194, 199–212
Sustainable education, x, 44, 46, 55, 57, 125
Systems leadership, 95, 98, 99, 102
Systems thinking, ix–x, 25–38, 92, 94–96, 99, 101, 102, 178, 180
Systems transformation, 65, 98, 101

T

21st century skills, 11, 13
2030 agenda, 85, 87, 88, 96–99, 102, 104, 124, 177, 205

U

United Nations leadership, 98
University-Industry Collaboration, xv, 166–168, 170–172, 174
University of the future, ix, xvi, 29, 38, 88–102, 139, 145
University transformation, 168
University with a soul, 121

V

Virtual reality (VR), 108, 109, 112, 113

Milton Keynes UK
Ingram Content Group UK Ltd.
UKHW022001140524
442584UK00001B/2